AMARANTH

Modern Prospects for an Ancient Crop

Report of an Ad Hoc Panel of the
Advisory Committee on Technology Innovation
Board on Science and Technology
for International Development
Office of International Affairs
National Research Council

This report has been reprinted by Rodale Press, Inc.,
with the permission of National Academy of Sciences.

NOTICE: The project that is the subject of this report was approved by the Governing Board of the National Research Council, whose members are drawn from the councils of the National Academy of Sciences, the National Academy of Engineering, and the Institute of Medicine. The members of the committee responsible for the report were chosen for their special competences and with regard for appropriate balance.

This report has been reviewed by a group other than the authors according to procedures approved by a Report Review Committee consisting of members of the National Academy of Sciences, the National Academy of Engineering, and the Institute of Medicine.

The National Research Council was established by the National Academy of Sciences in 1916 to associate the broad community of science and technology with the Academy's purposes of furthering knowledge and of advising the federal government. The Council operates in accordance with general policies determined by the Academy under the authority of its congressional charter of 1863, which establishes the Academy as a private, nonprofit, self-governing membership corporation. The Council has become the principal operating agency of both the National Academy of Sciences and the National Academy of Engineering in the conduct of their services to the government, the public, and the scientific and engineering communities. It is administered jointly by both Academies and the Institute of Medicine. The National Academy of Engineering and the Institute of Medicine were established in 1964 and 1970, respectively, under the charter of the National Academy of Sciences.

The Board on Science and Technology for International Development (BOSTID) of the Office of International Affairs addresses a range of issues arising from the ways in which science and technology in developing countries can stimulate and complement the complex processes of social and economic development. It oversees a broad program of bilateral workshops with scientific organizations in developing countries and conducts special studies. BOSTID's Advisory Committee on Technology Innovation publishes topical reviews of technical processes and biological resources of potential importance to developing countries. For a complete list of BOSTID publications, please write BOSTID (JH-217D), Office of International Affairs, National Research Council, 2101 Constitution Ave., Washington, DC 20418.

This report has been prepared by an ad hoc advisory panel of the Advisory Committee on Technology Innovation, Board on Science and Technology for International Development, Office of International Affairs, National Research Council. Funding for the study was provided by the William H. Donner Foundation and the Office of the Science Advisor, Agency for International Development, under Grant No. DAN/5538-G-SS-1023-00.

Art Credits:
Page IX depicts *Amaranthus caudatus* (from W.E. Safford in *Proceedings of the 19th Congress of Americanists, 1917*). The drawings on pages X and 54 depict various common shapes of amaranth.

Cover photograph by Burgess Blevins.

First printing, National Academy Press, September 1984
Second printing, Rodale Press, Inc., January 1985

ISBN 0-87857-560-X

Library of Congress Catalog Card Number 84-061583

PANEL ON AMARANTH

DONALD L. PLUCKNETT, Consultative Group on International Agricultural Research, The World Bank, Washington, D.C., *Chairman*

MELVIN G. BLASE, Department of Agricultural Economics, University of Missouri, Columbia

T. AUSTIN CAMPBELL, Economic Botany Laboratory, U.S. Department of Agriculture, Beltsville, Maryland

LAURIE B. FEINE, Rodale Research Center, Kutztown, Pennsylvania

HECTOR E. FLORES-MERINO, Yale University, New Haven, Connecticut

LINDA C. GILBERT, Product Development, Rodale Test Kitchen, Emmaus, Pennsylvania

RICHARD R. HARWOOD, Rodale Research Center, Kutztown, Pennsylvania

SUBODH JAIN, Department of Agronomy and Range Science, University of California, Davis

CHARLES S. KAUFFMAN, Rodale Research Center, Kutztown, Pennsylvania

CYRUS M. MCKELL, Native Plants Inc., Salt Lake City, Utah

GARY NABHAN, Native Seeds/SEARCH, Tucson, Arizona

HUGH POPENOE, International Programs in Agriculture, University of Florida, Gainesville

ALFREDO SANCHEZ-MARROQUIN, Instituto Nacional de Investigaciones Agricolas (Proyecto NAS/INIA), Mexico D.F.

ROBIN M. SAUNDERS, Western Regional Research Center, Cereals Unit, U.S. Department of Agriculture, Berkeley, California

JOSEPH SENFT, Amaranth Consultant, Emmaus, Pennsylvania

JAMES L. VETTER, American Institute of Baking, Manhattan, Kansas

DAVID E. WALSH, General Nutrition Corporation, Fargo, North Dakota

SPECIAL CONTRIBUTORS

G.J.H. GRUBBEN, Research Station for Arable Farming and Field Production of Vegetables, Lelystad, Holland

T.N. KHOSHOO, Secretary, Department of Environment, Government of India, New Delhi, India

JUDITH M. LYMAN, Rockefeller Foundation, New York, New York

JONATHAN SAUER, Department of Geography, University of California, Los Angeles

ARRIS A. SIGLE, Amaranth Farmer, Luray, Kansas

THEODORE SUDIA, U.S. Department of the Interior, Washington, D.C.

* * *

NOEL D. VIETMEYER, Professional Associate, Board on Science and Technology for International Development, *Amaranth Study Director*

National Research Council Staff

F.R. RUSKIN, *BOSTID Editor*
MARY JANE ENGQUIST, *Staff Associate*
CONSTANCE REGES, *Administrative Secretary*

Preface

Amaranth, a little-known crop of the Americas, is grown either as a grain crop or as a leafy vegetable. Despite its obscurity, it offers important promise for feeding the world's hungry. In the National Academy of Sciences' 1975 study *Underexploited Tropical Plants with Promising Economic Value,* amaranth was selected from among 36 of the world's most promising crops. Since then, extensive research has been done on the plant, and this book provides a more detailed examination of its characteristics and prospects.

The panel that produced this report met in September 1981 at the Rodale Research Center of Rodale Press in Emmaus, Pennsylvania. There, panel members examined a field of grain amaranth ready for harvest as well as test plots of several hundred amaranth varieties. They also sampled many amaranth products from the Rodale Test Kitchen. The panel members are indebted to Robert Rodale and his staff for their assistance and hospitality.

This report, resulting from the panel's deliberations, is intended for agencies engaged in development assistance and food relief, officials and institutions concerned with agriculture in developing countries, and scientific communities with relevant interests.

This study is one of a series that explores promising plant resources that heretofore have been unknown, neglected, or overlooked. Other titles include:

- *Underexploited Tropical Plants with Promising Economic Value* (1975)
- *Making Aquatic Weeds Useful: Some Perspectives for Developing Countries (1976)*
- *Tropical Legumes: Resources for the Future* (1979)
- *The Winged Bean: A High-Protein Crop for the Tropics* (second edition, 1981)

This series of reports is issued under the auspices of the Advisory Committee on Technology Innovation (ACTI) of the Board on Science and Technology for International Development (BOSTID), National Research Council. ACTI was established in 1971 especially to assess scientific and technological advances that might prove particularly applicable to problems of developing countries.

Funds for this study were provided by the William H. Donner Foundation, which also made possible the free distribution of the report. Staff support was provided by the Office of the Science Advisor, Development Support Bureau, Agency for International Development.

How to cite this report:
National Research Council. 1984. *Amaranth: Modern Prospects for an Ancient Crop*. National Academy Press, Washington, D.C.

Contents

AMARANTH

Modern Prospects for an Ancient Crop

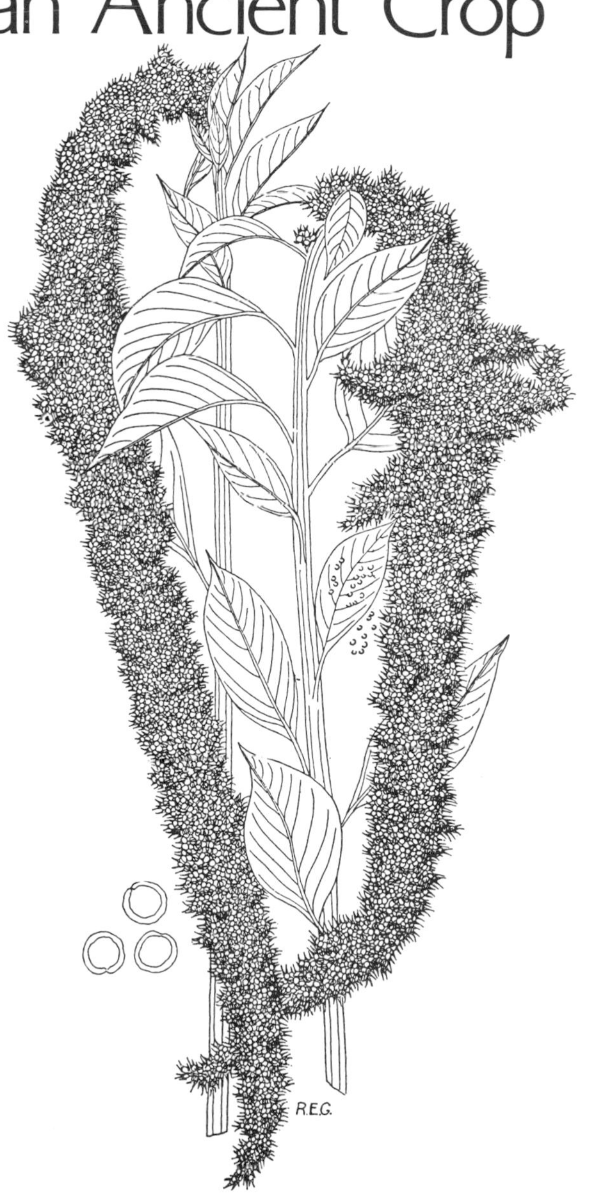

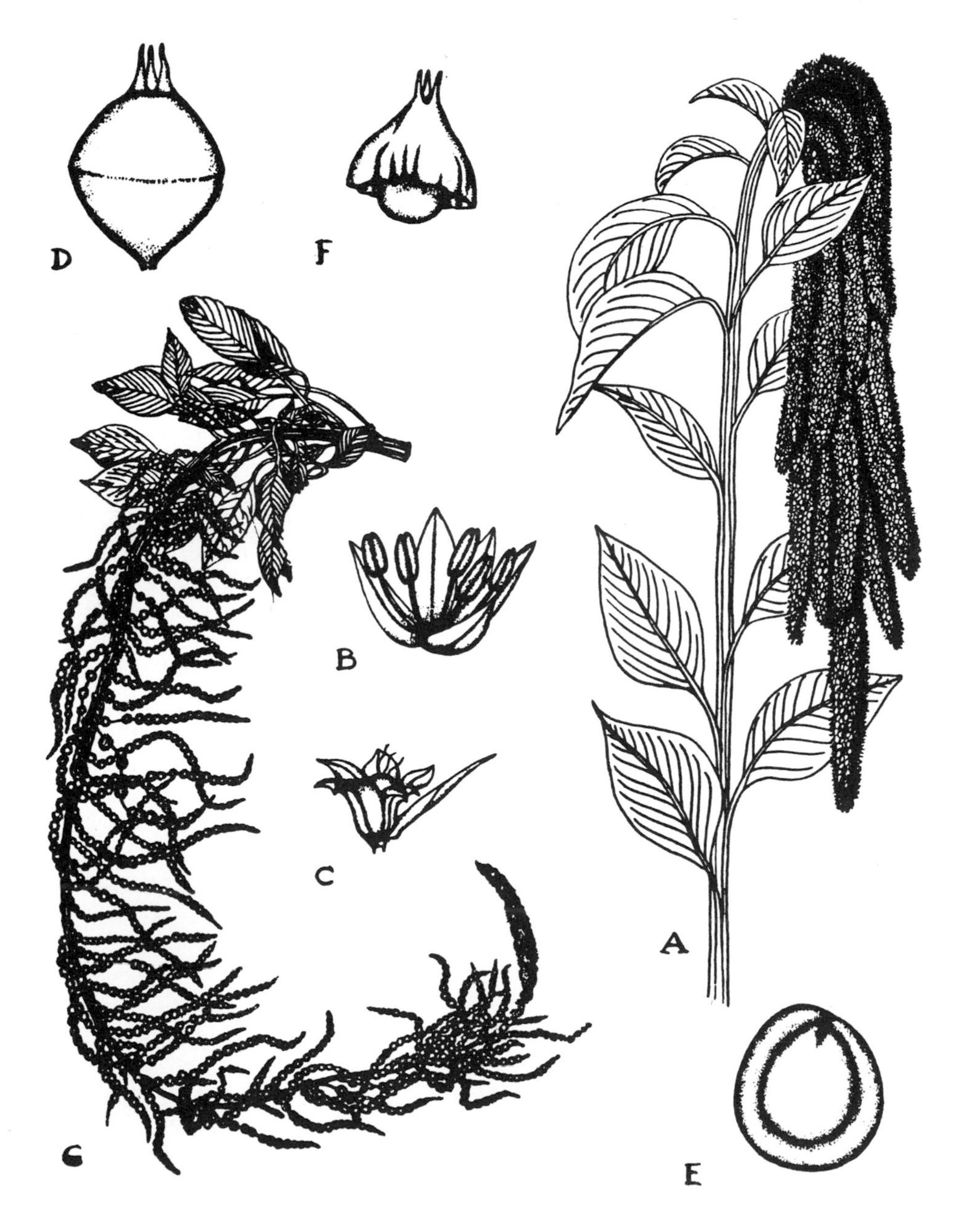
D
F
A
B
C
G
E

1

Introduction

In pre-Columbian times grain amaranth was one of the basic foods of the New World—nearly as important as corn and beans. Thousands of hectares of Aztec, Inca, and other farmland were planted to the tall, leafy, reddish plants. Some 20,000 tons of amaranth grain were sent from 17 provinces to Tenochtitlan (present-day Mexico City) in annual tribute to the Aztec emperor Montezuma.

Amaranth was interwoven with legend and ritual. On various days of the religious calendar, Aztec women ground the seed, mixed it with honey or with human blood, and shaped it into forms of snakes, birds, mountains, deer, and gods that were eaten either during ceremonies at the great temples or in little family gatherings.

Apparently, this use of amaranth in pagan rituals and human sacrifice shocked the Spanish conquistadors, and with the collapse of Indian cultures following the conquest, amaranth fell into disuse. In the Americas it survived only in small pockets of cultivation in scattered mountain areas of Mexico and the Andes. Corn and beans became two of the leading crops that feed the world, while grain amaranth faded into obscurity and today is largely forgotten. The Spanish conquest had ended amaranth's use as a staple of the New World and slowed the spread into world agriculture of a highly nutritious food. That amaranth was so important to the Aztec and other New World diets makes it a promising unconventional crop to investigate.*

Most of the world now receives the bulk of its calories and protein from a mere 20 species—notably cereals such as wheat, rice, maize,

*It is, however, only one of many underutilized food crops that are indigenous to Third World areas but neglected by researchers and policymakers. Others include the bambara groundnut *(Voandzeia subterranea)* and marama bean *(Tylosema esculentum)* of Africa; the winged bean *(Psophocarpus tetragonolobus)*, moth bean *(Vigna aconitifolia)*, and rice bean *(Vigna umbellata)* of Asia; quinua *(Chenopodium quinoa)* and tarwi *(Lupinus mutabilis)* of South America; and the tepary bean *(Phaseolus acutifolius)* of North America. These plants are described in the companion reports *Underexploited Tropical Plants with Promising Economic Value* (Report No. 16); *Tropical Legumes: Resources for the Future* (Report No. 25); and *The Winged Bean: A High-Protein Crop for the Tropics* (Report No. 37).

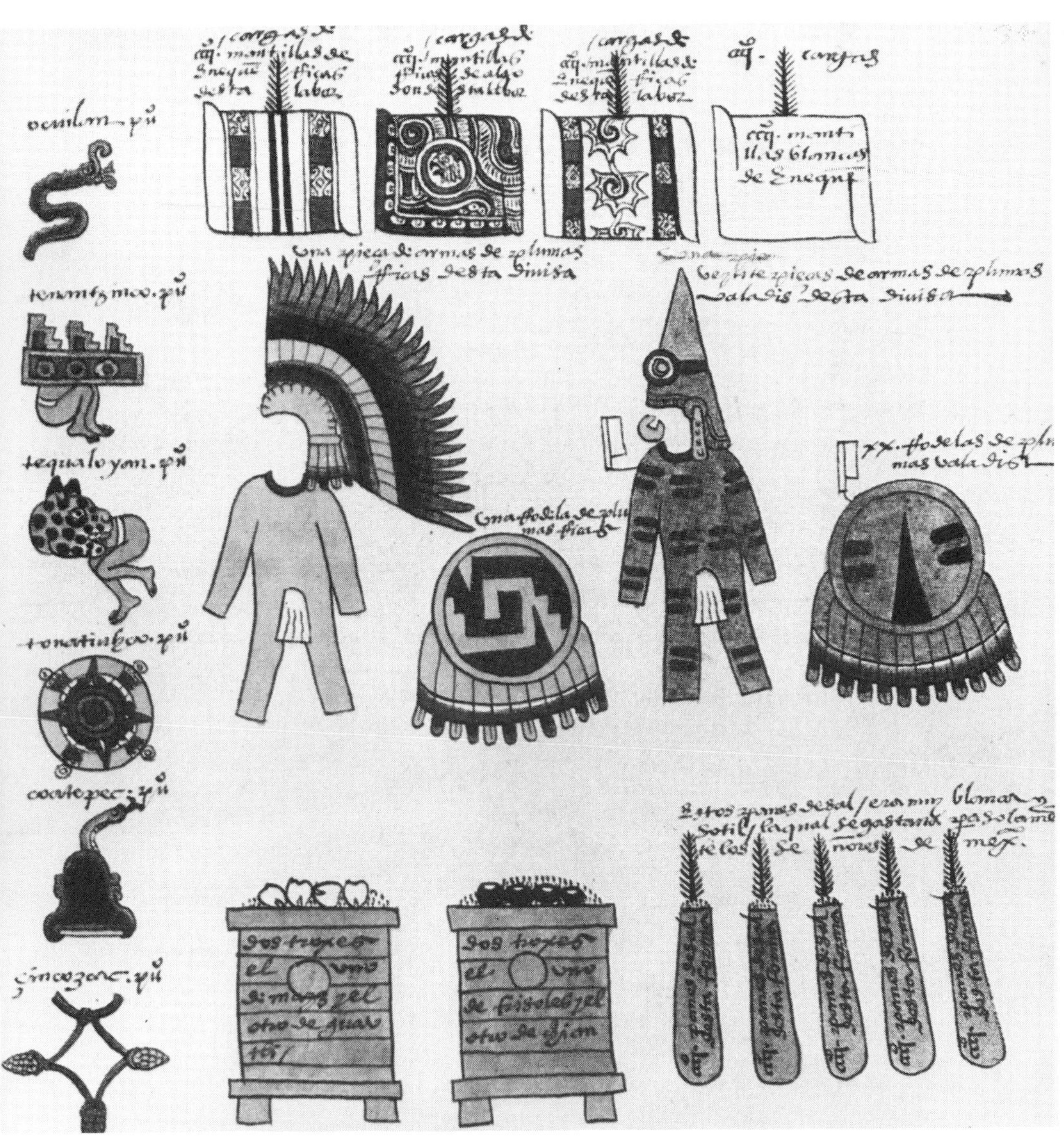

The Codex Mendoza, commissioned by the Spanish viceroy Antonio de Mendoza in about 1541, reveals that for two-thirds of the towns of the Aztec empire amaranth was a required part of the annual tribute paid to the emperor Montezuma II. The page shown above lists six towns in the left column and specifies their tribute: four bales of hennequen or cotton mantles in the designs shown; one war dress and shelf with rich feather trim; twenty war dresses with common feather trim; two wooden bins of maize and amaranth; two bins of beans and chia (another little-known Aztec crop); 2,000 salt loaves. Feathery symbols above drawings mean 400. (Drawing courtesy Organization of American States)

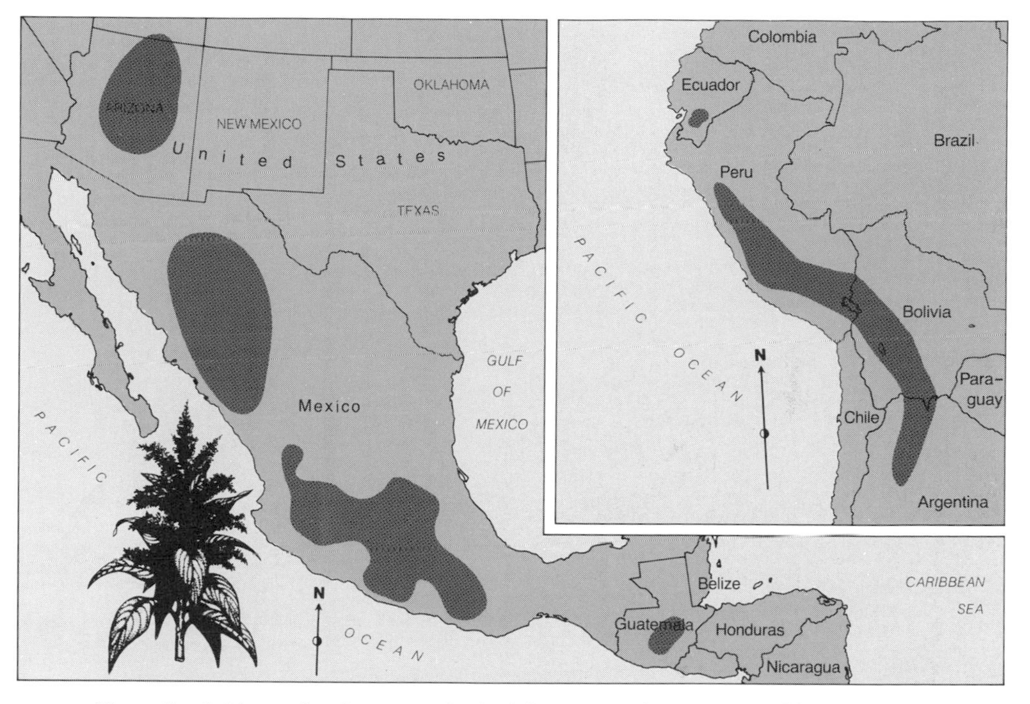

The native habitats of grain amaranths (mainly *Amaranthus cruentus* and *Amaranthus hypochondriacus*) are distributed throughout Mexico and extend into Guatemala and the southwestern United States. In South America (mainly *Amaranthus caudatus*) they are found in a band stretching from southern Ecuador through Peru and Bolivia into northern Argentina. (Adapted from information obtained from J. Sauer, courtesy *Encyclopaedia Britannica*)

millets, and sorghum; root crops such as potato, sweet potato, and cassava; legumes such as beans, peanuts (groundnuts), and soybeans; and sugarcane, sugar beet, and bananas. These plants are the main bulwark between mankind and starvation. It is a dangerously small larder from which to feed a planet.

To diversify the food base, we should not overlook lesser-known indigenous crops, such as amaranth. Some of them promise to become global resources. A century ago, the soybean, sunflower, and peanut were considered unworthy of concentrated research. Today, they are among the world's most important crops. Amaranth, too, could rise to universal prominence.

Grain Amaranths

Three species of the genus *Amaranthus* produce large seedheads loaded with edible seeds. *Amaranthus hypochondriacus and Amaranthus cruentus* are native to Mexico and Guatemala; *Amaranthus*

Amaranth is a broad-leafed plant but it produces an edible, cereal-like grain, as do grasses such as wheat, rice, rye, and barley. (Rodale Press, Inc.)

caudatus is native to Peru and other Andean countries. All three are still cultivated on a small scale in isolated mountain valleys of Mexico, Central America, and South America, where generations of farmers have continued to cultivate the crops of their forebears.

Amaranths are broad-leafed plants, one of the few nongrasses that produce significant amounts of edible "cereal" grain.* They grow vigorously; resist drought, heat, and pests; and adapt readily to new environments, including some that are inhospitable to conventional cereal crops.

Amaranth is a beautiful crop with brilliantly colored leaves, stems, and flowers of purple, orange, red, and gold.† The seedheads, some as long as 50 cm, resemble those of sorghum. The seeds, although barely bigger than a mustard seed (0.9–1.7 mm in diameter), occur in massive numbers—sometimes more than 50,000 to a plant—and are cream colored, golden, or pink.

With a protein content of about 16 percent, amaranth seed compares well with the conventional varieties of wheat (12–14 percent), rice (7–10 percent), maize (9–10 percent), and other widely consumed cereals. Amaranths began attracting increased research attention in 1972 when Australian plant physiologist John Downton found that the seed also contains protein of unusual quality. It is high in the amino acid lysine. Cereals are considered "unbalanced" in terms of amino acid composition because generally they lack sufficient amounts of lysine for optimum health. Amaranth protein, however, has nearly twice the lysine content of wheat protein, three times that of maize, and in fact as much as is found in milk—the standard of nutritional excellence. It is, therefore, a nutritional complement to conventional cereals. (Amaranth protein itself is low in leucine, but this amino acid is found in excess in conventional plant protein sources.)

Amaranth seed can be used in breakfast cereals or as an ingredient in confections. It also can be parched or cooked into gruel or milled to produce a sweet, light-colored flour suitable for biscuits, breads, cakes, and other baked goods. Amaranth grain, however, contains little functional gluten, so that it must be blended with wheat flour to make yeast-leavened baked goods "rise."

When heated, the tiny amaranth grains pop and taste like a nutty-flavored popcorn. The popped seeds are light and crisp and are eaten

*Others are quinua *(Chenopodium quinoa)* and buckwheat *(Fagopyrum* species). These three are sometimes called "pseudo cereals" to distinguish them from the grain-producing grasses.

†Several ornamental forms of these species are widely used all over the world; the western catalogs list types such as love-lies-bleeding and Prince of Wales feather, and one can easily find these and other amaranths, along with the closely related *Celosia* species, around houses, public parks, and government buildings.

In northern India amaranth grain is popped and mixed with honey to make a popular confection called laddoos. (M. Pal)

as a snack, as a cold cereal with milk and honey, as "breading" on meats or vegetables, or, held together with honey, as a sweet.

Although grain amaranth is a crop of the Americas, some time ago (probably since Columbus) *Amaranthus hypochondriacus* underwent a remarkable migration to Asia. There, during the last century, it became increasingly popular among hill tribes in India, Pakistan, Nepal, Tibet, and China. Travellers have reported seeing red and yellow patches of amaranths on the hillsides "tinging with flame the bare mountain slopes." In the Himalayas this amaranth species now is an important crop in a few local areas, and a flat bread made from its seeds is a common food. In fact, it is in this upland Asian region that this American crop now finds its most intensive cultivation. It often occupies more than half of the nonirrigated cropland of the higher elevations in the hills of Northwest India, for example.

Amaranth is gaining popularity also in the northwestern plains of India as well as in the hills of southern India under the common names rajgira ("king seed"), ramdana ("seed sent by God"), and keerai. Indians pop the grain and make it into confections (called laddoos) with honey or syrup, just the way the Aztec and Maya did centuries ago. And among Hindus, popped amaranth grain soaked in milk is now used on certain festival days when eating traditional cereals is forbidden.

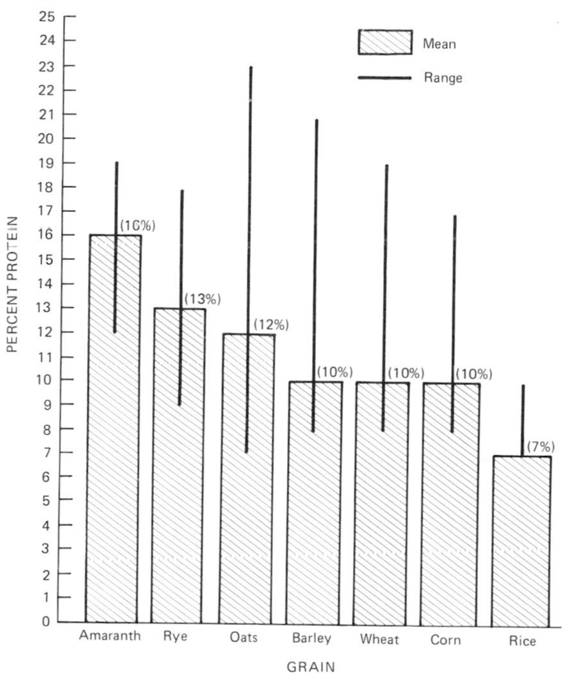

One exciting potential of amaranth grain lies in its high protein content as compared with other grains. . . .

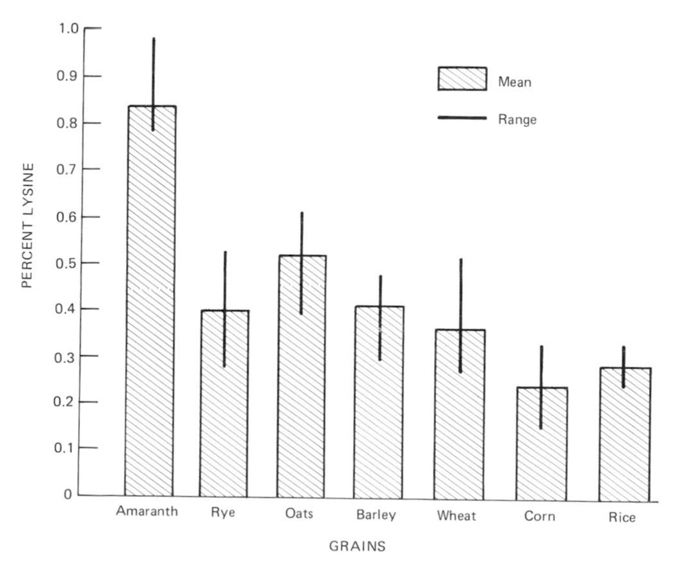

. . . Another virtue is the high amount of lysine and methionine, two nutritionally critical amino acids contained in amaranth protein. (Both are charts based on data collected by R. M. Saunders and R. Becker; see Tables 1 and 2, pages 35 and 36.)

Vegetable Amaranths

Seed is not the only nutritious product from the versatile amaranth. The leaves also are rich in protein as well as in vitamins and minerals. They have a mild flavor, and in much of the world young leaves and stems of amaranth are boiled as greens.* Although virtually unlisted in agricultural statistics, vegetable amaranths may actually be the most popularly grown vegetable crop in the tropics.

In the hot, humid regions of Africa, Southeast Asia (especially Malaysia and Indonesia), southern China, southern India, and the Caribbean, amaranth species such as *Amaranthus tricolor, Amaranthus dubius,* and *Amaranthus cruentus* are grown as soup vegetables or for boiled salad greens (potherbs). In North American deserts, where summers are too hot for lettuce or cabbage production, *Amaranthus palmeri* has long been a major wild green among Indians. In Greece, boiled *Amaranthus blitum* leaves have been a favorite salad (called vleeta) since the days of Homer.

But, as a recent U.S. Department of Agriculture bulletin pointed out,† few species of vegetables are so looked down upon. The demeaning phrase "not worth an amaranth" exists in several languages. Amaranths are sometimes thought fit only for pigs (hence the common name "pigweed" for one despised American species) and worthy of picking only when one is driven by poverty.

The same bulletin reports that:

> Among the vegetables of the tropics, few are as easy to grow as the amaranths. Starting from tiny seed, these species can produce delectable spinachlike greens in 5 weeks or less, can continue to produce a crop of edible leaves weekly for up to 6 months, and will then yield thousands of seeds to guarantee their survival. In favorable locations they can reseed themselves automatically and thus continue to produce a useful crop almost without attention. They can also invade plantings of other crops and become serious weeds, competing with slower growing weaker species for light and nourishment. In the tropics, amaranths can produce year-round. For little effort, they afford a nutritious dish with abundant provitamin A, a vitamin particularly necessary in the tropics for eye health. Amaranths also produce protein efficiently.

* In recent taste tests at the U.S. Department of Agriculture in Beltsville, Maryland, most of 60 participants said that cooked amaranth leaves tasted at least as good as spinach. Some likened the taste to that of artichoke.
† Martin and Telek, 1979.

Amaranth greens are a popular vegetable, grown throughout much of the tropics. Here in Benin, West Africa, a village woman uses palm leaves and a bowl of water to sprinkle the amaranth bed in her home garden. (G.J.H. Grubben)

Amaranth Development

Despite the growing evidence in amaranth's favor, much research needs to be done before the crop can be commercially produced. Agronomists are starting almost from scratch in adapting it to modern needs. Nevertheless, the researchers (see Appendix B) are learning the crop's responses to climate, soil conditions, pests, and diseases. Also, they are breeding short-statured plants of uniform height with sturdy, wind-resistant stalks and high-yielding seedheads that hold their seeds until they are harvested.

Leading amaranth's development is the Rodale Research Center near Emmaus, Pennsylvania, where more than a thousand different accessions collected from all parts of the world are being bred, grown, and evaluated. Further collaboration has been initiated with scientists in Africa, Asia, and Latin America; as a result, plant lines have been selected to overcome tendencies toward lodging, seed shattering, indeterminate growth, succulence at harvest time, and daylength dependence. This research effort has produced strains with improved baking, milling, popping, and taste qualities, as well as machinery adapted to planting, cultivating, harvesting, and threshing the crop. Lines of uniform color and height that bear their seedheads above the leaves, thus making them suitable for mechanical harvest, are now available. The crop can be said to be on the threshhold of limited commercial production in the United States, where twenty or so farmers are growing the crop. Several companies are testing the grain in their products, and an amaranth-based breakfast cereal is available.

Evidence indicates that amaranths adapt to many environments and tolerate adversity because they use an especially efficient type of photosynthesis to convert the raw materials of soil, sunlight, and water into plant tissues. Known technically as the C_4 carbon-fixation pathway, this process is used by a few other well-known fast-growing crops—sorghum, corn, and sugarcane, for example. The C_4 pathway is particularly efficient at high temperature, in bright sunlight, and under dry conditions. Plants that use it tend to require less water than the more common C_3 carbon-fixation pathway plants. For this reason, grain amaranth may be a promising crop for hot and dry regions.

Research has mainly emphasized grain amaranths so far, but in 1967 FAO started vegetable amaranth investigations. The following year it began field experiments in home garden projects in Nigeria and Benin. Later it commissioned germplasm collections. As a result, the vegetable branch of the amaranth family is beginning to attract recognition, and FAO has published a report on these species.*

*Grubben and Van Sloten, 1981.

Amaranth is a primitive crop that has received little modern research. However, a project in breeding and agronomic research at the Rodale Research Center in Pennsylvania has been remarkably successful at developing amaranth for modern farm conditions. Since 1976 more than 1,000 amaranths have been gathered from many parts of the world, and uniform varieties that can be mechanically sown, cultivated, and harvested have been developed. (Rodale Press, Inc.)

Cuzco, Peru. A research project is helping reestablish amaranth (''kiwicha'' in the Altiplano region of the Andes. University researchers have collected and genetically improved local varieties of *Amaranthus caudatus,* even to the point of producing fields of uniform, high-yielding plants that can be harvested by machine. In addition, they have perfected a flaked amaranth product with long shelf life, as well as a ''cannon'' that bursts amaranth seeds into a tasty puffed form. These developments have renewed local interest in amaranth, particularly among Indians, who now often come to the university to purchase foods such as their Inca ancestors lived on. (L. S. Kalinowski, Universidad Nacional de Cuzco)

YIELD

As yet, amaranth agronomists have paid little attention to improving seed yield. The plants are already quite productive, and researchers are concentrating on characteristics—such as ease of harvest and taste and food-processing qualities—that are more fundamental at this early stage. In Pennsylvania, test plots of productive strains of grain amaranths routinely yield 1,800 kg of seed per hectare. In California and elsewhere, small trial plots have yielded up to nearly twice that amount, and at four locations in the Himachal Pradesh and Uttar Pradesh hills of India, lines selected from the local land races have yielded 3,000 kg of grain per hectare.* Therefore, researchers suspect that grain amaranths will eventually match the yields of most other cereals.

*Information from B.D. Joshi.

As with any new crop, there are many production uncertainties. Problems reported at the various research locations include the lygus bug (an insect that sucks nutrients out of the immature seeds), fast-growing weeds that in some environments overwhelm the slow-starting amaranth seedling, and strains of amaranth that produce seedheads so heavy that they flop over during a summer thunderstorm. These, however, are not insurmountable difficulties.

FUTURE PROSPECTS

Together, the grain- and vegetable-type amaranths could provide many nutritious foods for the world. The small seed size is a limitation in planting as well as in harvesting, threshing, and cleaning the grain. But modern experience in the northern Indian plains shows that they have a good chance of adapting successfully and of being quickly accepted by villagers in Third World areas where they are adapted.

In the farming of the future, amaranths could find several valuable niches. They might complement other cereals such as sorghum, millets, or barley, thus helping countries that import large amounts of wheat by providing a locally grown extender. In addition, they would provide a local source of feed grain for the burgeoning poultry industries of developing nations. And, in particular, grain amaranth is a promising new crop for dry lands (areas with 600–800 mm of rainfall per year), for tropical highlands up to extreme elevations (3,500 m and above), and as a quick-maturing, dry-season crop for monsoon areas.

It would be a mistake, however, to expect amaranth to be on dinner plates next year; it took a century for the American public and the farmers to accept the soybean, and it took two centuries for Europeans to accept the potato. Compared with such now-established crops, amaranth has been the object of little modern research or testing. Nevertheless, with today's communications and technology, it should not take that long for amaranth to find its niche. Within a few years, it seems likely that this ancient grain of the Americas will return to grace the modern age. Eventually, it may prove to be as rich a legacy of the American Indian as maize and beans.

2

The Plants

Members of the genus *Amaranthus* (family Amaranthaceae) are widely distributed throughout the world's tropical, subtropical, and temperate regions. The genus contains about 60 species.* Growth habits vary from prostrate to erect and branched to unbranched; leaf and stem colors range from red to green, with a multitude of intermediates; and seed colors range from black to white.

DISTRIBUTION AND ECOLOGY

Historically, people have cultivated amaranths in environments ranging from the true tropics to semiarid lands and from sea level to some of the highest farms in the world. Ecotypes have evolved that tolerate alkaline sandy soils with pH as high as 8.5, as well as the acidic clays of hillside slash-and-burn fields of the tropics.

Although traditionally cultivated within 30° latitude of the equator, amaranth can be grown in higher latitudes using strains that will initiate flowering in spite of the longer daylength (photoperiod) than that of the tropics. Most grain amaranth cultivation has been concentrated in highland valleys, such as those in the Sierra Madre, Andes, and Himalayas. Generally, traditional farmers in those areas have exchanged seed only with neighbors, so that adjacent semiarid, highland, and subtropical lowland gene pools have remained largely distinct.

PHYSIOLOGY

Amaranths, as already noted, are among the group of plants that carry on photosynthesis by the specialized C_4 pathway. They are one of the few C_4 crop species that are not grasses.

*Willis, J.C. 1973. A Dictionary of the Flowering Plants and Ferns. Cambridge University Press.

The C_4 pathway is a modification of the normal photosynthetic process that makes efficient use of the carbon dioxide available in the air by concentrating it in the chloroplasts of specialized cells surrounding the leaf vascular bundles. The photorespiratory loss of carbon dioxide, the basic unit for carbohydrate production, is suppressed in C_4 plants. Consequently, plants that use the C_4 pathway can convert a higher ratio of atmospheric carbon to plant sugars per unit of water lost than those possessing the classical C_3 (Calvin cycle) pathway.

Even when their stomata are partially closed, plants having the C_4 pathway are able to maintain relatively high rates of carbon dioxide fixation. Since stomata close when the plant is under environmental stress (such as drought or salinity), C_4 plants, such as amaranth, perform better than C_3 plants under adverse conditions. Also, the reduced stomatal opening reduces the water lost by transpiration.

Through osmotic adjustment, the plants can tolerate some lack of water without wilting or dying. This is also an adaptation for surviving periods of drought.* The potential ability to photosynthesize at high rates under high temperature is another physiological advantage of C_4 photosynthesis. Research on *Amaranthus caudatus* cv *"edulis"* has shown peak photosynthetic activity to occur at 40°C.†

DAYLENGTH

Many of the amaranths are sensitive to length of day. For example, strains of *Amaranthus hypochondriacus* from the south of Mexico will not set flower in the summer in Pennsylvania. They do, however, mature in the greenhouse during the short-day conditions of winter. The reverse happens with *Amaranthus cruentus* from Nigeria. It remains vegetative for a long period in its equatorial home. However, it goes to seed very early when introduced into the long-day conditions in Pennsylvania and can be used to breed for early-maturing traits. *Amaranthus caudatus,* on the other hand, is known to be a short-day species. It usually flowers and sets seed only when daylength is less than 8 hours.‡ However, some *Amaranthus caudatus,* such as the ornamental "love-lies-bleeding," will set seed at longer day-length conditions.

*Information from J. Ehleringer.
†El-Sharkawy et al., 1968.
‡ Fuller, 1949.

ALTITUDE

Elevation is not a severe limitation. Amaranths grow satisfactorily from sea level to above 3,200 m, but only *Amaranthus caudatus* is known to thrive at altitudes above 2,500 m.

TEMPERATURE

Grain amaranth grows best when the daily high temperature is at least 21°C. Various accessions have showed optimal germination temperatures varying between 16°C and 35°C. The speed of emergence is increased at the upper end of this range.

Although *Amaranthus hypochondriacus* and *Amaranthus cruentus* tolerate high temperatures, they are not frost hardy. Growth ceases altogether at about 8°C and the plants are injured by temperatures below 4°C. *Amaranthus caudatus,* however, is native to areas high in the Peruvian Andes and is more resistant to chilling than the other species.

SOIL

Field observations indicate that amaranth grows well on soils containing widely varying levels of soil nutrients. Initial studies in Pennsylvania show that young grain amaranth plants grow taller with fertilizer, but the grain yield has thus far shown little improvement. Vegetable amaranth, on the other hand, requires high soil fertility, particularly potassium and nitrogen.

Grain amaranth requires well-drained sites and appears to prefer neutral or basic soils (pH values above 6). However, this has not been studied carefully, and, with the wealth of amaranth germplasm that exists, it is likely that types that tolerate acid conditions can be found, especially as vegetable amaranth is often cultivated in tropical lowlands where acid soils are common.

Although the genus is not known for high salt tolerance, an apparent ability to withstand mild salinity and alkalinity is apparent in some species of amaranth. Moreover, *Amaranthus tricolor* has demonstrated tolerance for soil with high aluminum levels.*

RAINFALL

For seeds to germinate and establish roots, amaranths require well-moistened soil, but once seedlings are established, grain amaranths do

*Foy and Campbell, 1981

well with limited water; in fact, they grow best under dry, warm conditions. Vegetable amaranths, on the other hand, require moisture throughout the growing season. Grain amaranths have been grown in dry-land agriculture in areas receiving as little as 200 mm of annual precipitation, and, at the other extreme, vegetable amaranths are routinely grown in areas receiving 3,000 mm of annual rainfall. Indeed, in West Africa, vegetable amaranth production continues even during the torrential rainy season.

WEEDY AMARANTHS

Of all the 60 or so amaranth species, only a handful are now used as crops. A few of the others are serious weeds. The main weedy types are *A. viridis, A. spinosis, A. retroflexus,* and *A. hybridus*. These are dark-seeded plants with widespread distribution. *A. retroflexus* ("pigweed") is one of the world's worst weeds.

These weeds are distinctly different from the grain amaranths highlighted in this report. They tend to be indeterminate and to produce seeds at many different parts of the plant and scatter them during a long part of the season. On the other hand, the cultivated grain types tend to mature over a rather short period and have a dominant seedhead with fewer side branches.

Like other *Amaranth* species, the hardy weeds prefer hot, bright sunlight. Their seed is readily spread by birds and water. They frequently occur as pests in pastures, crops, or along roadsides, usually in unshaded areas in competition with other weeds and grasses. In urban areas they are commonly seen in abandoned lots or poking up through cracks in the pavement.

The seeds of weedy amaranths have remarkably long viability; some have germinated after 40 years. (Much less is known about the viability of the seed of cultivated amaranths, except that seeds stored at room temperature in desiccators may keep well for several years. In high humidity the seeds quickly lose viability, and they also lose their popping quality with age or desiccation.)

Of the weedy species, only some types of *A. hybridus* are worth serious attention as vegetable crops or as breeding parents for grain amaranth improvement.

The species and varieties cultivated for grain do not carry the same potential for weediness. They lack the strong taproot that is seen in *A. retroflexus,* for example, and are generally much less aggressive in their habits.

3

Production

PLANTING

Amaranths are usually seeded directly into the field. On rare occasion, vegetable amaranths are transplanted to the field as seedlings when they reach the stage of bearing four true leaves. Seeding density depends on the method of harvest anticipated. Preliminary density trials indicate that, for many *Amaranthus hypochondriacus* and *Amaranthus cruentus* cultivars, 320,000 plants per hectare is an acceptable density for yield as well as for stand management. Vegetable amaranths are often grown with densities up to 100 plants per m^2.

Seedbeds should be of good tilth, well drained, and fairly level to help prevent rain from washing away the tiny seeds or seedlings. Seeds must be planted no more than 1 cm deep, and the seedbeds must have fine soil without large clods. Rain falling on this type of seedbed can cause the soil to form a crust that inhibits germination.

CULTIVATION

Once the stand is established, maintenance is relatively easy. The broad leaves and erect habit quickly create a closed canopy, making understory weeds only a minor problem under most conditions.

Grain amaranths can be mechanically weeded until the canopy closes. Most types of grain amaranth mature in 4–5 months. They mature more quickly in monsoonal areas. However, in some highland regions maturing may take as long as 10 months.

HARVESTING

A few varieties selected in Pennsylvania are now sufficiently uniform to be machine harvested, but most are not. The main difficulty in mechanical harvesting is that the central flower head matures and dries out while the numerous inflorescences on lower side branches are still moist. High-density planting modifies plant structure to a point where

Field of maturing amaranth, Emmaus, Pa. (Rodale Press, Inc.)

a single seedhead is formed, and this makes mechanical harvest more efficient. Selection and breeding of new varieties that are adapted to mechanical harvest is now in progress.

To avoid the problem of nonuniform maturity, the grain amaranth now produced in Latin America and South Asia is hand harvested. It is then dried in the sun and threshed and winnowed by hand. The small seed size makes cleaning awkward. However, winnowing is not too difficult, since the seed is heavier than the chaff.

5 4

Amaranth farmers, traditional and modern. 1. Northern India (M. Pal); 2. Bhutan (R.P. Croston, IBPGR); 3. the high Andes of Peru (C.S. Kauffman); 4. Kansas, USA (Rodale Press, Inc.); and 5. Mexico (L. Feine).

2

3

PESTS

Insect problems are not well documented. In Pennsylvania the lygus bug *(Lygus lineolaris)* has severely damaged grain amaranth yields by piercing the developing seed and sucking out the juices. *Amaranthus cruentus* (white-seeded types) seem more susceptible to damage than *Amaranthus hypochondriacus*. Leaf-miners have also been found on both grain and vegetable amaranth. In Lucknow, India, serious damage to both grain and vegetable types has often been caused by spider mites. In India the stem weevil *(Lixus truncatulus)* is a major pest of amaranth; its grubs damage foliage and roots and cause the plants to wilt. Leaf rotters *(Hymenia recurvalis)* also cause considerable damage during rainy seasons.

Seedbeds should be guarded against ants and termites, which often carry away all the seeds.

DISEASES

More study is needed of diseases on both vegetable- and grain-type amaranths. The soil fungus *Alternaria alternantherae* causes much leaf damage and drastically reduces plant vigor. However, only *Amaranthus caudatus* has so far proved susceptible to this disease in Pennsylvania. A similar blight disease of leaves and flowers, caused by *Alternaria amaranthi* (Peck.), has been reported on *Amaranthus hypochondriacus* in India. White rust, caused by *Albiyo bliti,* in which white pustules on the underside of the leaves reduce the market appeal of vegetable amaranth, is also common in South India.

There is evidence that a foliar disease (or perhaps air pollution) affects *Amaranthus cruentus* in the United States. Leaf and stem diseases caused by mycoplasms have been identified in Peru, and great differences in susceptibility between lines have been noted.

Opposite
Amaranth is a strikingly colored crop. Clockwise from top, *Amaranthus caudatus, Amaranthus cruentus, Amaranthus hypochondriacus* (all Rodale Press, Inc.), and an ornamental variety developed in India at the Lucknow Botanical Gardens for the brilliance of its leaves (T.N. Khoshoo).

Overleaf
Amaranth farmer and his crop in the uplands of western Nepal. (© Tony Hagen, courtesy National Geographic Society)

Page 26
Amaranth seed is extremely small. It varies in color from black (vegetable type) to golden (grain type), and heating causes it to pop to form a light, white product (shown at the bottom). (Rodale Press, Inc.)

The major promise of amaranth grain is its blends with cereals such as wheat. Incorporated into breads and other baked goods, amaranth flour improves their nutritional quality, particularly because it supplements the amino acids they lack. (Rodale Press, Inc.)

4

Grain Amaranths

As already mentioned, amaranth is one of the few nongrasses with potential for becoming a cereal-like grain crop. The main species for this are *Amaranthus caudatus*, *Amaranthus cruentus*, and *Amaranthus hypochondriacus*.

AMARANTHUS CAUDATUS

This species is a crop in the Andean highlands of Argentina, Peru, and Bolivia. It has pendulous, blazing-red inflorescences and is commonly sold in Europe and North America as an ornamental under names such as "love-lies-bleeding" or "red-hot cattail," a name shared with unrelated plants. Other forms of the species give much better grain yields. One good variety that has club-shaped inflorescences is *Amaranthus caudatus* cv *edulis* (sometimes classified as *Amaranthus edulis* or *Amaranthus mantegazzianus*).

Amaranthus caudatus originated in the same region in the Andean highlands as the common potato. The Spanish conquerors called it Inca wheat, but it is much more ancient than the Incas. Some of its pale seeds, placed in tombs as food for the dead, are more than 2,000 years old.

The plant is still widely grown in the Andean region, mostly by the Indians who maintain traditional customs. It is usually planted in small patches close to houses, not in large fields as a staple crop. The grain is toasted and popped, ground into flour, or boiled for gruel. It is considered especially good for children and invalids.

The crop contains a great deal of genetic diversity in South America, and, although only a small sampling has been introduced to other continents, much genetic diversity has been observed in the germplasm collections from northern India.

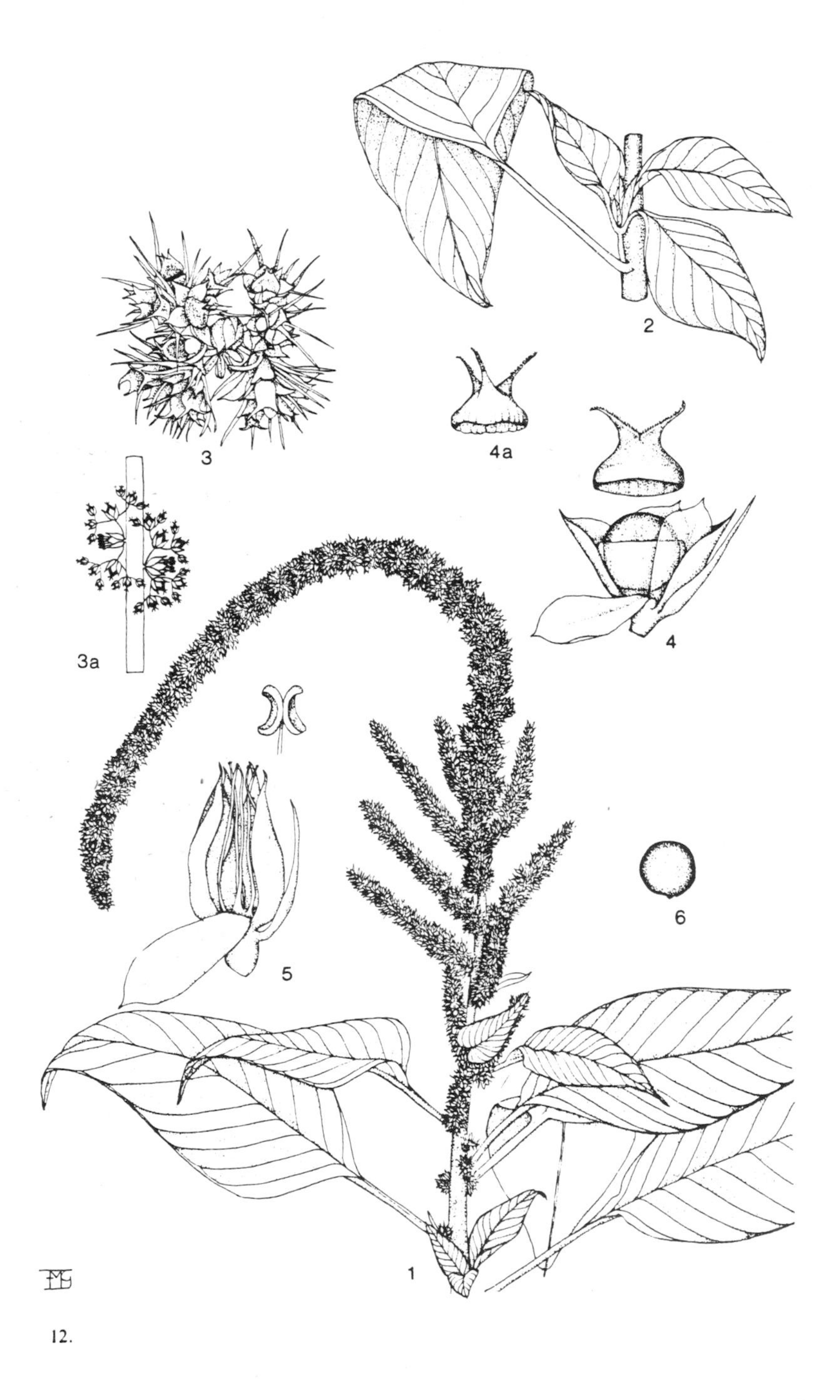

Amaranthus caudatus (Line drawings on pages 28, 30, 42, and 44 are reprinted with permission of the Department of Plant Taxonomy and Plant Geography, Wageningen Agricultural University, Wageningen, The Netherlands. For complete taxonomic details of numbered items, see Grubben and van Sloten, 1981.)

AMARANTHUS CRUENTUS

This Mexican and Guatemalan species is useful both as a grain or a leafy vegetable (see next chapter). The grain types have white seeds; the vegetable types (as well as those used to extract red dye) usually are dark seeded. It is probably the most adaptable of all amaranth species, and it flowers, for example, under a wider range of daylengths than the others.

Amaranthus cruentus is an ancient food, and in the famous Tehuacan caves in central Mexico, archaeologists have dug up remains—both the pale grain and the bundles of plants brought in for threshing—at a dozen levels, dating back 5,500 years. The species is still grown in the region, and popped amaranth seedcakes are sold on the streets of the towns. *Amaranthus cruentus* has also survived as a grain crop in a few Indian villages of southern Mexico and Guatemala and as a crop used to extract a red dye for coloring corn-based foods in the Indian pueblos of the arid southwestern United States, where it probably became established in prehistoric times.

AMARANTHUS HYPOCHONDRIACUS

The most robust, highest yielding of the grain types, *Amaranthus hypochondriacus* was probably domesticated in central Mexico, farther north and at a later time than *Amaranthus cruentus*. It first appeared in the Tehuacan caves about 1,500 years ago as a pale-seeded, fully domesticated type. It, too, reached the United States in prehistoric time but later became extinct there. Its maximum cultivation today is in India, particularly in the Sutlej Valley in the state of Himachal Pradesh and in the Garhwal and Kumaon regions of Uttar Pradesh.

Some types of *Amaranthus hypochondriacus* are bushy; others are tall and unbranched. The species is particularly useful for tropical areas, high altitudes, and dry conditions. It has excellent seed quality and shows the greatest potential for use as a food ingredient. It pops and mills well and has a pleasing taste and smell.

Evidently, the Spanish took seeds back to Europe at an early date (possibly inadvertently), and, as shown by sixteenth- and seventeenth-century herbals, the plant soon spread through European gardens as an ornamental. Around 1700 it was grown as a minor grain crop in Central Europe and Russia and eaten as mush and groats. By the early nineteenth century it had been taken to Africa and Asia, where it is now planted as a grain crop in such widely scattered regions as the mountains of Ethiopia, the hills of South India, the Nepal Himalaya, and the plains of Mongolia.*

*Although the plant is undoubtedly of American origin, just when and how it was introduced to Asia is not clear. It is cultivated in remote and isolated areas of the eastern Himalayas, where even potato was introduced only in recent decades. This, along with the ethnobotanical and genetic evidence, gives the impression that it is an ancient and well-established crop. (Information from T.N. Khoshoo)

Amaranthus cruentus

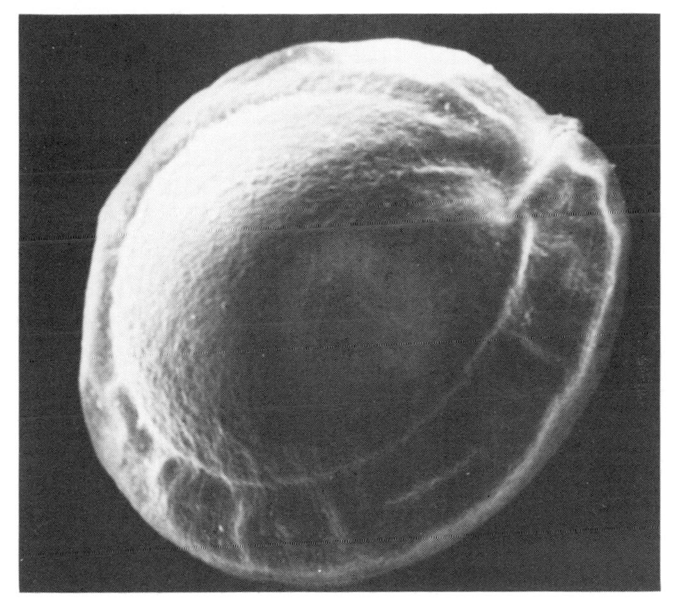

Under magnification an amaranth seed looks strikingly like a flying saucer. (R. Saunders)

PHYSICAL COMPOSITION

Amaranth seeds are very small; 1,000–3,000 seeds per gram are common. Although selections have been made over the years for pale seeds (the wild species all have black seeds), large inflorescences, and more seeds per plant, there has apparently been little selection for larger seed size.

Amaranth germ and bran constitute 26 percent of the seed; the flour 74 percent—about the same as in a grain of wheat. When the whole grain is milled, its protein, vitamins, fat, and minerals are concentrated significantly in the bran/germ fraction. Amaranth germ, for example, can contain as much as 30 percent protein. It also contains about 20 percent oil. This shows promise as an edible oil, but no attempts at extracting it have yet been made. The bran is high in fiber, protein, vitamins, and minerals.

The starch that makes up the bulk of amaranth flour has extremely small granules (average diameter 1 micron*), a unique dodecahedral structure, and high water-absorption capacity. It is likely to prove useful for applications in the food, plastics, cosmetics, and other industries.

*Information from L. St. Lawrence, G.J.L. Griffin, and R. Saunders.

Algería, a confection of the Aztecs, is still popular in Mexico and Central America. To make it, the mature amaranth plants are harvested (1) and the seed shaken out (2). The seed is popped on a hot stone or metal sheet over a fire (3). The popped product is mixed with honey or molasses (4), and compressed into blocks for sale (5). (1 and 2, L. Feine; remainder, Rodale Press, Inc.)

4

5

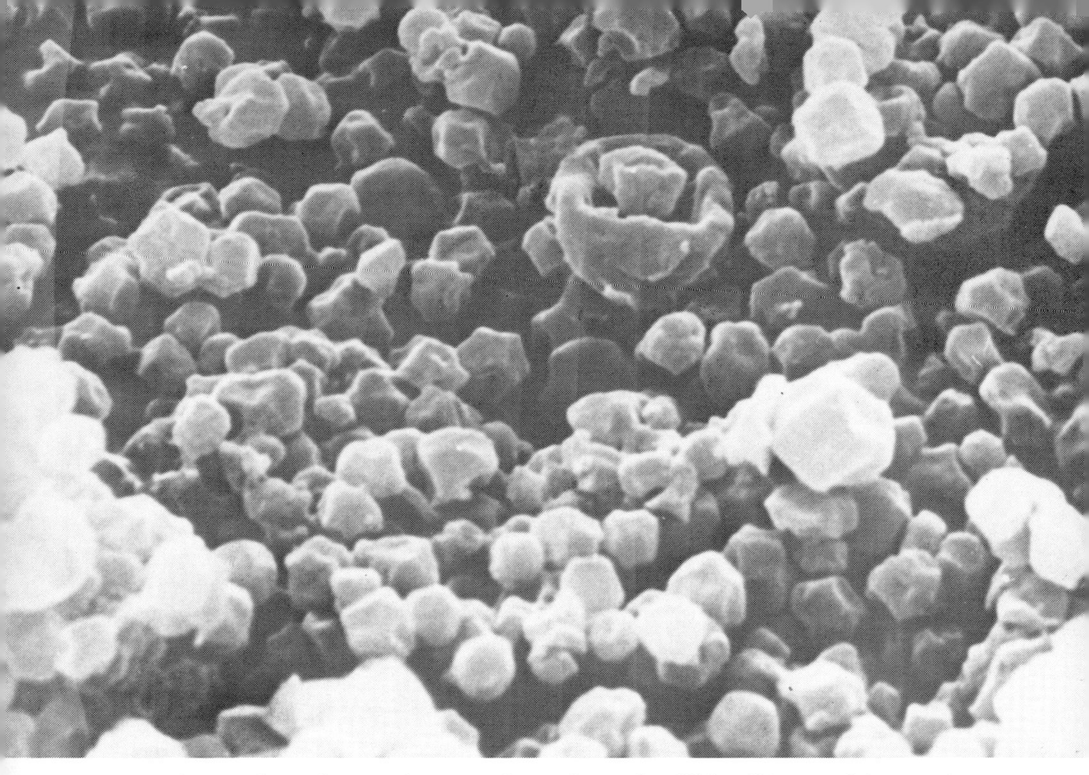

Amaranth starch comes in extremely small granules. With a diameter of about 1 micron, they are among the smallest ever recorded. This is a potentially important commercial advantage. The granules also have an almost crystalline structure, an extremely unusual characteristic. (L. St. Lawrence and G.J.L. Griffin)

CHEMICAL COMPOSITION

Average chemical composition of amaranth grain is shown in Table 1, amino acid content in Table 2. Amaranth has a protein content as high as 16 percent, which is somewhat higher than that found among commercial varieties of common cereals. However, the "white" flour that is milled out of it has only 7 percent protein, not a substantially different quantity from the protein content of wheat flour used in making white bread.*

The charts on page 7 show the composition of amaranth in comparison with the more common grains. The protein in amaranth seeds is unusual because its balance of amino acids is closer to the optimum balance required in the human diet than that of most plant proteins. As noted already, the lysine content is especially high compared with that found among the most common cereals. Thus, amaranth's importance is that its essential amino acids complement those of corn, rice, and wheat. For example, corn protein is low in both tryptophan and lysine, whereas amaranth has high levels of both.

*Information from R. Saunders.

TABLE 1 Proximate Composition of *Amaranthus* Seeds[a]

Species	N (%)	Crude Protein[b] (%)	Fat (%)	Fiber (%)	Ash (%)
A. cruentus[c,d]	3.05	17.8	7.9	4.4	3.3
A. cruentus × *hypochondriacus*[d]	2.97	17.4	8.0	4.3	3.0
A. edulis[d] (*A. caudatus*)	2.70	15.8	8.1	3.2	3.2
A. hypochondriacus[e,f]	2.67	15.6	6.1	5.0	3.3

[a] Dry basis (original moisture contents, 6–11%)
[b] N × 5.85
[c] Average of two *A. cruentus* samples
[d] Becker et al., 1981
[e] Cheeke and Bronson, 1980
[f] Average of four *A. hypochondriacus* samples

Source: Saunders and Becker, 1983

Amaranth protein, itself, is low in the amino acid leucine, which is not a serious limitation because leucine is found in excess in most common grains.

The nutritional value of amaranth protein is very good. Protein efficiency ratios (PER) have ranged from 1.5 to 2.0 (corrected to casein 2.5) for cooked grain, and its total digestibility is about 90 percent. Amaranth protein, at a biological value of 75, comes closer than any other grain protein to the perfect balance of essential amino acids, which theoretically would score 100 on the nutritionists' scale of protein quality based on amino acid composition. By contrast, corn scores about 44, wheat 60, soybean 68, and cow's milk 72. When amaranth flour is mixed with corn flour, the combination almost reaches the perfect 100 score, because the amino acids that are deficient in one are abundant in the other.

The fatty acids of amaranth oil comprise about 70 percent oleic and linoleic acids, about 20 percent stearic acid, and about 1 percent linolenic acid. The oil also contains uncommonly high levels of squalene.

Antinutritive factors, such as saponins, trypsin inhibitors, and tannins, occur in amaranth grain but at similar levels to those found in legumes and in some other grains (notably sorghum). Much more information is needed on these components, but as of now they are not thought to present any nutritional hazard.

PROCESSING

As with other grains, foods containing amaranth can be prepared by using simple low-energy techniques. Cleaned, unprocessed whole grain can be made into porridge by simply boiling it briefly in water. If toasted or parched lightly, the whole grain becomes a pleasant-tasting food that can be eaten without further preparation. The whole

TABLE 2 Protein Amino Acid Composition (g/16 g of N) of *Amaranthus* Species

Amino Acid	Species		
	A. caudatus[a]	*A. hypochondriacus*[a]	*A. cruentus*[b]
Lysine	5.3	5.5	5.1
Histidine	2.5	2.5	2.4
Threonine	3.5	3.6	3.4
Cysteine	2.3	2.1	2.1
Methionine	2.4	2.6	1.9
Met + Cys	4.7	4.7	4.0
Valine	4.1	4.5	4.2
Isoleucine	3.6	3.9	3.6
Leucine	5.3	5.7	5.1
Tyrosine	2.8	3.3	2.6
Phenylalanine	3.4	4.0	3.4
Serine	5.9	6.3	5.4
Glycine	6.9	7.4	7.0
Arginine	. . .	. . .	7.9
Alanine	. . .	. . .	3.4
Aspartic acid	. . .	. . .	7.8
Glutamic acid	. . .	. . .	14.2
Proline	. . .	. . .	3.6
Tryptophan	. . .	. . .	. . .
Nitrogen recovered	89.6	86.8	85
Chemical score[c]	75	81	73

[a] Carlsson, 1980
[b] Betschart et al., 1981
[c] The chemical score of whole wheat protein is 73, and of soybean protein, 74 (FAO, 1970)

Source: Saunders and Becker, 1983

grain can also be sprouted for use as a nutritious vegetable. Moreover, tasty foods can be prepared by popping or puffing the whole grain into small white kernels that taste like popcorn. This preparation is common in Mexico and Central America, where popped amaranth is often used in confections and condiments.

Grinding or milling amaranth produces a whole-grain meal or white flour. An abrasive mill like that used for milling sorghum or rice seems best, although a roller mill, as used for wheat, can be adapted to handle amaranth. As with other grains, milled products have a much shorter storage life than the whole grain.

USE IN FOODS

Amaranth meal or flour is especially suitable for unleavened (flat) breads where it can be used as the sole or predominant cereal ingredient. The flour is used in Latin America and in the Himalayas to produce a variety of flat breads (for example, tortillas and chapaties).

For making yeast-raised breads or other leavened foods, amaranth meal or flour must be blended with wheat meal or wheat flour because it lacks functional gluten. In such blends, it is likely that the high

Individual seedheads may contain tens of thousands of the exceedingly small amaranth grains. (N.D. Vietmeyer)

lysine content of amaranth improves the protein quality of foods that normally would be made from flours of other grains such as corn, rice, or wheat. This is particularly beneficial for infants, children, and pregnant and lactating women.

Amaranth can be used in many other foods, including:

- Soups (grain and flour)
- Pilaf (grain)
- Pancakes (flour, whole grain, and popped grain)
- Breakfast cereals (whole, popped, or sprouted-grain flour)
- Porridge (popped grains in milk)
- Breads, rolls, muffins, and many other forms of baked foods (flour, popped grain, toasted grain, whole grain)
- Crêpes (flour, popped grain)

- Dumplings, tostadas, tortillas, fritos, and corn pones (flour, whole or popped grain)
- Cookies and crackers (flour, whole or popped grain)
- Snack bars (popped grain, toasted grain, or sprouted grain)
- Toppings (popped grain)
- Breadings (popped grain, flour)
- Beverages (flour, popped grain)
- Fillers (whole or popped grain, flour, or starch)
- Confections (popped grains).

FEED

Unprocessed amaranth grain probably can be used as an animal feed, particularly for poultry. So far, however, it has not been produced in sufficient quantities to be tested extensively in livestock feeding trials.

5

Vegetable Amaranths

Most *Amaranthus* species have edible leaves, and several species are already widely used as potherbs (boiled greens). Over the years, growers have selected types with leaves and stems of high palatability. Their mild spinach-like flavor, high yields, ability to grow in hot weather, and high nutritive value have made them popular vegetable crops, perhaps the most widely eaten vegetables in the humid tropics. In some African societies, for example, protein from amaranth leaves provides as much as 25 percent of the daily protein intake during the harvest season.

In many tropical regions of India, China, Southeast Asia, and the South Pacific islands, amaranths, such as *Amaranthus tricolor* and *Amaranthus dubius,* are grown. They have been cultivated for more than 2,000 years, and many different cultivars, some with brightly colored leaves, have been developed. In humid tropical Africa, *Amaranthus cruentus* is extensively grown as a leaf vegetable. In much of the Caribbean, *Amaranthus dubius* and other amaranths are considered among the best of leafy greens.

In temperate regions of Eurasia, amaranths have long been domesticated as leafy vegetables, as were their relatives, spinach and chard. *Amaranthus lividus*, a low-growing, succulent, purplish-red herb, was cultivated in the gardens of ancient Greece and Rome and in medieval Europe. In North American deserts, Indians subsisted on *Amaranthus palmeri* and *Amaranthus hybridus* until corn and beans could be harvested. They were the only summer leafy vegetables with a dependable yield in the hot and arid conditions.

Despite the fact that vegetable amaranths are cultivated or gathered in so many regions, few references include more than generalities about their culture. This may indicate both their ease of cultivation and the fact that, because of wide adaptability, the optimal conditions for maximum yields are not known. Often the seeds are not sown at all and natural seed fall provides the following year's crop.

Since the plants grow rapidly, the time between planting and harvest of the tender foliage and stems is short—generally only 3–6 weeks. In

Chiang Mai, Thailand. Wild amaranth leaves collected for human and pig food. (M.G.C.McD. Dow)

Tamil Nadu (South India) plants are pulled 3 weeks after sowing and used as "tender greens." Certain varieties remain succulent longer and can be harvested after up to 5 weeks of growth. Varieties suitable for periodical cutting (clipping) are also available. The first cutting is done at 20 days after sowing, and thereafter weekly cuttings are possible for up to 10 cuttings. At a later age, and particularly after the flowers develop, the foliage and stems become fibrous, brittle, pithy, and unpalatable.

AMARANTHUS CRUENTUS

This species has been described earlier as a grain type. A very deep-red, dark-seeded form of the species, sometimes known as blood amaranth, is often sold as an ornamental in commercial seed packets. During the nineteenth century, this deep-red form was adopted for use as cooked greens by gardeners throughout the tropics. It became a more important crop in tropical Africa than anywhere else. Like corn, sweet potatoes, peanuts, and other American Indian crops, *Amaranthus cruentus* was evidently introduced to Africa by Europeans. But then it passed quickly from tribe to tribe, probably as a weed in millet and sorghum seed. It outran European exploration of the interior, so that Livingstone and others found it already in cultivation when they arrived. Today it is being planted and gathered year-round in the humid regions of much of Africa. In parts of West Africa, for instance, the tender young seedlings are pulled up by the roots and sold in town markets by the thousands of tons every year.

AMARANTHUS DUBIUS

This weedy species is used as a green vegetable in West Africa and the Caribbean and is found in Java and other parts of Indonesia as a home garden crop. One of the best varieties of this species, known as the cultivar "claroen," is particularly popular in Benin and Suriname. Its seeds are extremely small (4,500 seeds per g). It has distinctive dark-green, broad, ridged leaves. It is fast growing, high yielding, and has considerable morphological variation, resulting partly from its repeated hybridization with *Amaranthus spinosus*. The "greens" of this species are considered very palatable. This is the only tetraploid (2n = 64) species in the genus known so far.*

AMARANTHUS HYBRIDUS

This weedy species is one of the most common leafy vegetables. It is an herb that grows up to 1.5 m tall. Originating in tropical America, it is now spread throughout tropical areas. In Indonesia it is often planted in kitchen gardens. It also grows wild on moist ground, in waste places, or along roadsides, and thrives in altitudes up to 1,300 m.

In the markets, bundles of leafy shoots as well as uprooted young plants are offered for sale. The tender leaves and young seedlings are used widely in soups and stews. Lamb, beef, chicken, or pork is often added.

The common name of the plant in Mexico is quintonil. Elsewhere it is known as phak khom (Thailand), bayam (Malaysia, Indonesia), urai (Philippines), and slender amaranth.

The size and color vary greatly. Red-stemmed varieties are usually planted as ornamentals; green varieties are commonly used as vegetables.

Amaranthus hybridus has potential to impart early maturity to grain types through crossbreeding.

*Information from T.N. Khoshoo.

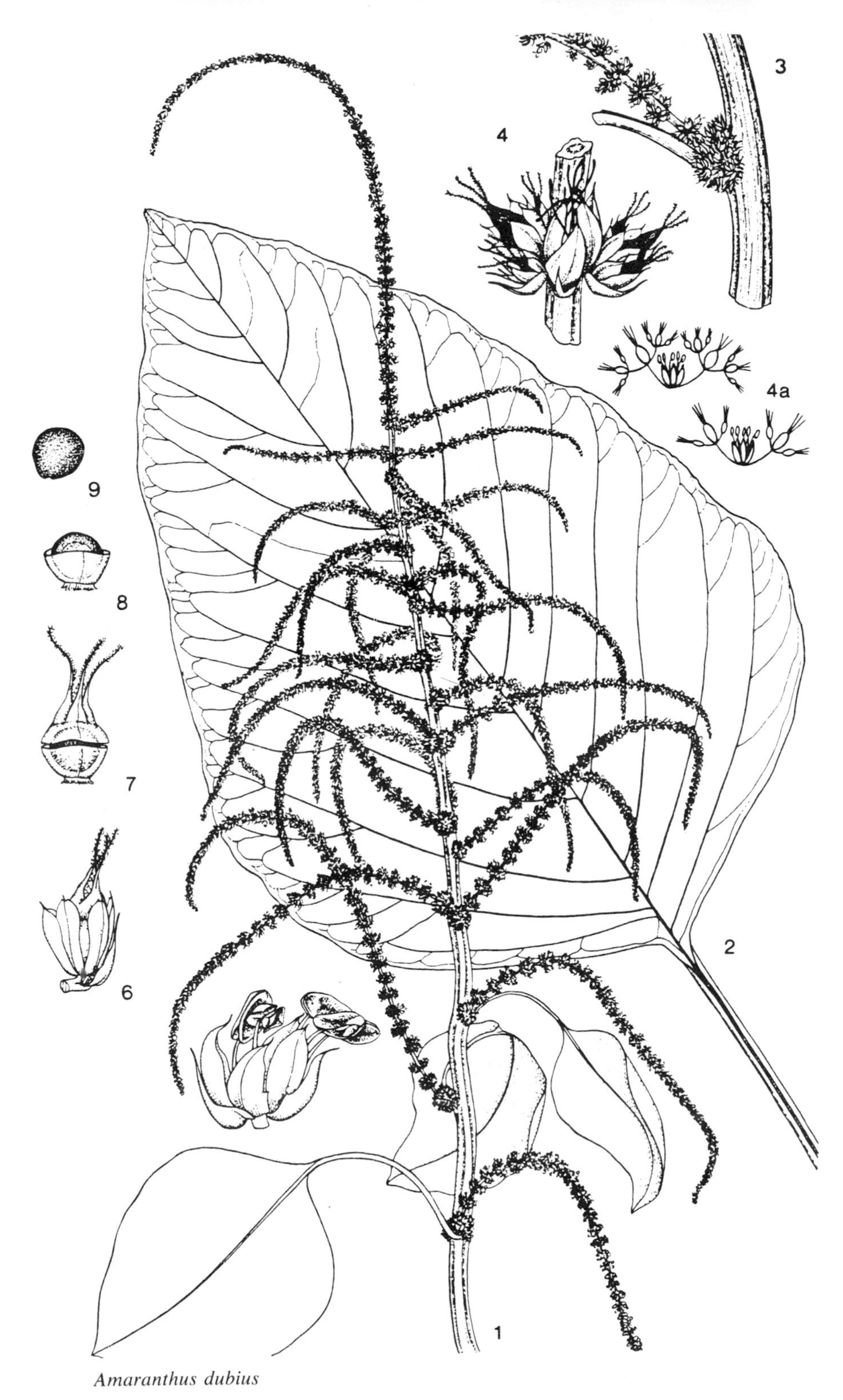

Amaranthus dubius

AMARANTHUS LIVIDUS

This widely distributed species (also known as *A. blitum*) is well adapted to temperate climates and has a number of weedy forms that come with either red or green leaves. It promises to allow the development of highly palatable crossbred vegetable amaranths. In Madhya Pradesh, India, the edible forms, known as norpa, are especially liked for their tender stems. This is the species widely eaten in Greece under the name vleeta. It is also grown in Taiwan, where it is known as horsetooth amaranth.

AMARANTHUS TRICOLOR

Varieties of this species are native to a large area from India to the islands of the Pacific and as far north as China. It is probably the best-developed of the vegetable amaranth species. The plants are succulent, low growing, and compact, with growth habits much like spinach. They can be produced as a hot-season leafy vegetable in arid regions when few other leafy greens are available. In India a number of domesticated forms are available, especially in Andhra Pradesh, Karnataka, Tamil Nadu, and Kerala. Some ornamentals with very beautiful foliage also belong to this species.

NUTRITIONAL QUALITY

The nutritional quality of amaranth greens is similar to that of other leafy vegetables. However, because their dry-matter content is often high, an equivalent amount of fresh amaranth often provides from 2 to 3 times the amount of nutrients found in other vegetables (see Table 3). In mineral content, notably iron and calcium, amaranth greens rank particularly well when measured against other potherbs.

Leaf-protein levels (dry-weight basis) have been reported as 27 percent for *Amaranthus blitum,* 28 percent for *Amaranthus hybridus,* 30 percent for *Amaranthus caudatus,* and 33 percent for *Amaranthus tricolor.** The amino acid composition of *Amaranthus hybridus* leaf protein shows a chemical score of 71, which is comparable to that of spinach.

High levels of the nutritionally critical amino acids lysine and methionine have been found in the leaves of 13 amaranth species.† Vegetable amaranths are also an important source of vitamins, especially vitamin A, the lack of which results in a most serious nutritional deficiency in the tropics and leads to blindness in thousands of children each year.

*Information from Monika Ipanez S.
†Koch et al., 1965.

Amaranthus tricolor

TABLE 3 Nutrient Content[a] of Selected Raw Vegetable Leaves[b]

Component	Amaranth	Spinach	Malabar spinach *(Basella)*	Chard
Dry matter, g	13.1	9.3	6.9	8.9
Food energy, cal	36	26	19	25
Protein, g	3.5	3.2	1.8	2.4
Fat, g	0.5	0.3	0.3	0.3
Carbohydrates				
Total, g	6.5	4.3	3.4	4.6
Fiber, g	1.3	0.6	0.7	0.8
Ash, g	2.6	1.5	1.4	1.6
Calcium, mg	267	93	109	88
Phosphorus, mg	67	51	52	39
Iron, mg	3.9	3.1	1.2	3.2
Sodium, mg	. . .	71	. . .	14.7
Potassium, mg	411	470	. . .	550
Vitamin A, IU	6,100	8,100	8,000	6,500
Thiamin, mg	0.08	0.10	0.05	0.06
Riboflavin, mg	0.16	0.20	. . .	0.17
Niacin, mg	1.4	0.6	0.5	0.5
Vitamin C, mg	80	51	102	32

[a] Per 100 g of edible portion
[b] From Watt and Merrill, 1963

Source: Saunders and Becker, 1983

YIELD

Generally, yields are in the range of 4 to 14 tons per ha green weight.* However, vegetable amaranth yields have been reported as high as 40 tons per ha. Fertilization, especially with nitrogen, is one of the major factors influencing yield, although few, if any, fertility trials have been done and there is little data for different growing regimens or locales.

Regrowth can provide four or more harvests a year. In Benin, it has been suggested that the plants be cut when they reach a height of 20 cm and that the harvest interval should be 3 weeks.†

DISEASES AND PESTS

Insect and disease problems can seriously affect vegetable amaranths. The plants do not grow well during long periods of cloudy, wet weather and are not tolerant of shade. During a monsoon rainy season, for example, damping-off from *Pythium* and *Rhizoctonia* is most serious. Another disease, *Choanephora cucurbitarum*, causes wet rot in leaves and young stalks.

To reduce such fungal diseases, seedbeds (especially those estab-

*Campbell and Abbott, 1982.
†Grubben, 1980.

lished during humid weather) must be well drained and located in sunny sites. Manuring can eliminate some of the problems. Various fungicides have also been used successfully.

Insect damage can be a more serious problem for vegetable amaranths than for grain amaranths. The major pests are larvae of moths and butterflies, as well as leaf hoppers, leaf miners, grasshoppers, and leaf-feeding beetles.

Slugs and snails also often severely damage young plantings.

LIMITATIONS

Green leafy vegetables are known to contain a wide variety of antinutritional factors. Amaranth is no exception. Oxalic acid, beta-cyanins, alkaloids such as betaine, cyanogenic compounds, saponins, sesquiterpenes, and polyphenols all have been reported present in various species.

Like many other fast-growing plants, amaranth requires and absorbs large amounts of nitrate. Under certain conditions, nitrate accumulates to levels that constitute several percent of plant dry matter. The levels, however, are about the same as those in spinach, beet greens, chard, and other conventional potherbs. Boiling the leaves removes most of the nitrate.

Green leafy vegetables, including amaranth, also frequently contain large quantities of oxalic acid. Oxalic acid is an end product of metabolic processes and accumulates as the plant gets older. When eaten by humans, it binds some minerals, notably calcium, and makes these unavailable for absorption from the digestive tract, so that a person consuming sufficiently large amounts of them develops a mineral deficiency.

The oxalic acid levels in amaranth samples can be uncomfortably high (1–2 percent), particularly when the plant is grown under dry conditions. The levels are affected by soil fertility and increase with fertilization. It is because of oxalic acid that amaranths (and many other leafy greens) are boiled before eating. (Boiling makes amaranths nontoxic, because the oxalic acid dissolves in the water.)

FORAGE

The fact that boiled amaranth leaves have been an important component of the human diet in many African and southern Asian countries for centuries suggests that it might also be a useful forage crop for animals, particularly ruminants.

Extensive studies to confirm this have not been conducted, and there are a number of reports in the veterinary literature that implicate species of wild amaranth in the poisoning of livestock. However, these incidents were under dry conditions where weedy amaranths are known to accumulate high levels of nitrate and oxalates and where they survive better than other plants, so the animals could not balance their diet. These observations raise some concern over the possible usefulness of amaranth as a forage crop; research to resolve the issue is well warranted.

LEAF-PROTEIN ISOLATES*

A future promise of vegetable amaranths is the development of leaf-protein concentrates. Compared with most other species, amaranth leaf protein is highly extractable. In one trial, amaranth had the highest level of extractable protein among 24 plant species studied. During the extraction of leaf protein, most other nutrients are extracted as well: for example, provitamin A (beta-carotene), polyunsaturated lipids (linoleic acid), and iron. Heating or treating the extract with acid precipitates the nutrients as leaf-protein concentrate. In the process, most of the harmful compounds are eliminated, as they remain in the soluble phase. The green cheeselike coagulum is washed with water slightly acidified with dilute acetic acid (vinegar) to further reduce the amounts of possible antinutritive factors. The resulting leaf-nutrient concentrate is especially useful for young children and other persons with particularly high protein, vitamin A, and iron needs. The fibrous pulp left after extracting the amaranth greens is a suitable feed for animals.

The protein quality of the amaranth leaf-nutrient concentrate (determined by amino acid composition, digestibility, and nutritional effectiveness) is excellent. It is, however, species dependent, probably because of the presence of secondary substances in some species.

*Information from R. Carlsson.

6

Research Needs

As noted, amaranth research has been gaining momentum in recent years. While no fundamental obstacles to the crop's future development are apparent, many technical details remain to be explored. Agencies funding agricultural research for developing countries should consider supporting amaranth research and testing. Some recommendations and research needs are listed here.

COLLECTION AND SCREENING OF GERMPLASM

Amaranth offers more genetic diversity in its present undeveloped state than do many conventional crops. The broad geographic spread of the genus has resulted in the evolution of many land races in widely separated areas. Several features of their highly variable breeding system and overall reproductive biology provide ample choice of breeding methods.

This huge gene pool will be very important to the future development of the crop. The genetic characteristics of the types to be found there should be useful in amaranth-breeding programs worldwide, and it is best to have as many available as possible at this early stage in amaranth development. Further systematic collections of amaranth germplasm should be made in Latin America, the Caribbean, India, Nepal, China, and the Pacific.

These collections should be coordinated with the International Board on Plant Genetic Resources (IBPGR), which is starting germplasm collections in Southeast Asia and has recently completed one in Peru.

ADAPTABILITY TRIALS

Although a large number of accessions and progeny from breeding selections have already been screened in Pennsylvania, there has been no organized program to determine the relative adaptability of the better varieties elsewhere. Therefore, it is recommended that a cooperative research network be established in which perhaps 12 promising amaranth varieties are grown in 12 or more locations in different climates.

The varieties selected for adaptability trials should include a range of species and morphological types that have agronomic potential. This would be an excellent way to obtain indications of the geographical areas where given types will grow best. It would be the first step towards developing "zones" for amaranth adaptability, similar to those used for soybeans.

Larger trials, involving perhaps 100–500 varieties, should also be undertaken by geneticists and screening nurseries. Demonstration trials under farm conditions are also recommended.

ETHNOBOTANY OF THE CROP

Studies of the way people grow and use grain amaranths in Central and South America and the Himalayas, as well as those methods employed to grow and use vegetable amaranths elsewhere, could prove informative. Data could be gathered about the most favorable environments within which the various species can be grown. And much useful information could be collected by documenting the rainfall, temperature, daylength, and soil conditions of these areas. This would permit better targeting to new areas. It would also demonstrate:

- Lesser-known uses;
- Soil requirements;
- Crop ecology (for example, insect damage); and
- Social aspects of amaranth cultivation and use.

Although such studies will benefit those wanting to grow amaranth for the first time, it will also assist people who traditionally cultivate the crop.

WEEDY AMARANTHS

Seeds of the shattering, weedy amaranth species (for example, *Amaranthus hybridus, Amaranthus palmeri, Amaranthus retroflexus,* and *Amaranthus spinosus*) should *not* be distributed for cultivation. Nevertheless, these weeds can be useful to the amaranth breeder. *Amaranthus hybridus* is the wild progenitor species of present-day cultivated *Amaranthus hypochondriacus* and easily exchanges genes with it. For introducing such desirable traits as faster maturation, disease resistance, and wider adaptability in the cultivated forms, the material of *Amaranthus hybridus* could be invaluable. And *Amaranthus spinosus* (a diploid), though sometimes a troublesome weed, can be utilized in raising hybrids (F_1 triploids) with *Amaranthus dubius* (a tetraploid) on a commerical scale for forage. This is made possible by the peculiar distribution of male and female flowers in *Amaranthus spinosus*. The hybrids are fast growing, sterile, and have very soft spines. Feeding trials and nutritional studies are, however, a prerequisite before using such hybrids for forage.*

AMARANTH TAXONOMY

A taxonomic "key" to amaranths compiled by Laurie Feine is available.† However, because amaranth taxonomy is so confused, a monograph on the family is needed.

SOME SPECIFIC RESEARCH NEEDS

GRAIN AMARANTH GROWING PRACTICES

Amaranth cultivation and harvesting practices require several types of research. Many of the following items are site-specific.

- Selection of types best adapted to local conditions.
- Use of amaranth in mixed-cropping systems and rotation.
- Adaptation—it is important to learn the most extreme rainfall, evaporation, and soil characteristics under which amaranth can be grown and produce a reasonable yield.
- Soil requirements—fertility, tolerance for salinity, need for organic matter, and maximum and minimum moisture.
- Effects of growth conditions on chemical composition and nutritive value.
- Determination of best planting dates, plant density, and weed and pest management.
- Development or adaptation of machinery: planting implements, thresher, winnower, and grain cleaner.
- Control of diseases (for example, pythium and rhizoctonia fungi) and pests (such as lygus bug).

*Information from M. Pal.
†Grubben and van Sloten, 1981.

• Storage of seeds (for example, the development of techniques to prevent "caking up," which makes planting more difficult).

DEVELOPING GRAIN AMARANTH VARIETIES

Breeding goals for grain amaranth include selection for:

• Desirable growth characteristics such as reduced plant size, reduced sensitivity to photoperiod, synchronous flowering, early maturity, reduced lodging, uniform drydown, reduced shattering, increased seed size, and high yield;
• Environmental adaptations such as drought tolerance, pest and disease resistance, herbicide tolerance, and efficient fertilizer utilization; and
• Food quality such as white seeds, palatability, and high levels of protein and essential amino acids.

GRAIN AMARANTH PROCESSING

Research is needed on the physiology of postharvest handling, especially on the effect of moisture on grain quality and storage. Also needed are studies on:

Amaranth tissue culture. As of 1984, amaranth can be cultured from cells. This advance should greatly accelerate and facilitate the breeding of amaranths. (J. Ehleringer and NPI, Inc.)

- Grain cleaning (especially for medium-scale pilot plants);
- Removal of sand and weed seeds from the grain;
- Adaptation of existing machinery to handle amaranth; and
- Drying and storage of harvested grain (unthreshed grain is hygroscopic).

Research is also needed on:

- Processing the whole grain, such as by extrusion cooking, milling whole and popped seed, and toasting, rolling, sprouting, and popping. Any changes in nutritional value or chemical compounds from such processing need to be assessed.
- Commercial development requirements, such as dry and wet milling, and derivatives (for instance, methylated and carboxy derivatives of amaranth starch).
- Storage and shelf life of products.
- Value of the grain and crop residues for ensilage and for direct feeding to livestock.

GRAIN AMARANTH FOOD USES

Important research needs in the use of amaranth grain include:

- Basic characteristics of seed starch, protein, bran, germ, and oil;
- Uses in products, including breakfast foods and weaning mixtures, as well as recipe development;
- Nutritional testing in humans;
- Amaranth's value as a wheat extender or a supplement for added nutritional value in traditional foods such as chapaties, tortillas, weaning foods, chicha, and arepas;
- Use in infant foods;
- Nutritional availability of minerals, vitamins, protein, and starch;
- Amaranth's functional characteristics (viscosity, density, freeze-thaw stability, heat stability, emulsifying properties) when used in foods, and how grain types differ from one another; and
- Antinutritional constituents.

VEGETABLE AMARANTHS

Vegetable amaranths have been more thoroughly investigated than the grain amaranths. Selections have been made by Asian growers for many years. Named varieties suitable for widespread culture are available from seed companies in Hong Kong, Taiwan, and the United States (see Appendix C). Nevertheless, the crop could be improved by studies of:

- Pest and disease resistance;
- Nutrient uptake and nutrient content at different stages of harvest or crop growth;
- Leaf yield;
- Food quality, including tenderness and storage methods to prolong the life of the harvested produce;
- Use of amaranth leaves as a remedy for vitamin A deficiency;
- Antinutrition factors and heavy-metal accumulation in response to type and quantity of fertilizers used and type of soil;
- Production of leaf nutrient concentrate;
- Regrowth after harvest;
- Comparison of yield from clipping versus successive planting;
- Seed production and farmer-selection techniques;
- Leaf:stem ratio;
- Late emergence of inflorescences;
- Planting and cultural practices for efficient use of land, water, and fertilizer; and
- Crop rotation to avoid soil-borne diseases.

The benefits and possible toxic problems of vegetable amaranth as a forage also need study.

NEW USES

It is possible that amaranth products other than those described in this report could have important potential. Examples are:

- Natural dyes;
- Pharmaceuticals (for example, laxatives); and
- Squalene (a high-priced material found in amaranth seed but normally obtained from shark livers and used in cosmetics).

ENVIRONMENTAL EFFECTS

It is important to study the weediness of the most promising *Amaranthus* species and the likelihood of their becoming pests.

It is also possible that amaranth pollen and grain may cause undue allergic reactions in some people, and this needs to be assessed.

Amaranthus tricolor

Appendix A

Selected Readings

An excellent bibliography of amaranth literature has been collected and compiled by the Rodale Research Center. (Senft, J. P., C. S. Kauffman, and N. N. Bailey. 1981. The Genus Amaranthus: A Comprehensive Bibliography. Rodale Press, Inc., Emmaus, Pennsylvania, USA. 217 pp.) A most useful publication for libraries and for those interested in amaranth research, it contains more than 2,600 entries and references organized according to the following subjects: general; genetics/taxonomy; weeds; horticulture/agronomy; plant physiology; plant pathology/entomology; anthropology; and nutritional/food utilization.

An amaranth newsletter is being published to help readers keep current with the latest research developments. It includes abstracts of amaranth publications. For further information contact: Editor-in-Chief, *Archivos Latinoamericanos de Nutricion,* P.O. Box 1188, Guatemala City, Guatemala, C.A.

Allen, P. 1961. Die Amaranthaceen Mitteleuropas. Die Naturlichen Pflanzenfamilien. Reprint from G. Hegi, Illustriert Flora von Mitteleuropa III 2:461–535, Munich, Germany.

Aguilar, J., and G. Alatorre. 1978. Monografia de la planta de la alegría. Memoria 1978. Grupo de Estudios Ambientales 1(1):156–203.

Becker, R., E. L. Wheeler, K. Lorenz, A. E. Stafford, O. K. Grosjean, A. A. Betschart, and R. M. Saunders. 1981. A compositional study of amaranth grain. Journal of Food Science 46:1175–1180.

Bosworth, S. C., C. S. Hoveland, G. A. Buchanan, and W. B. Anthony. 1980. Forage quality of selected warm season weed species. Agronomy Journal 72(6):1050–1054.

Brenan, J. P. M. 1981. The genus *Amaranthus* in Southern Africa. Journal of South African Botany, Pretoria 47(3):451–492.

Campbell, T. A., and J. A. Abbott. 1982. Field evaluation of vegetable amaranth (*Amaranthus* spp.). HortScience 17(3):407–409.

Carlsson, R. 1982. Leaf protein concentrates from plant sources in temperate regions. Pp. 52–80 in Leaf Protein Concentrates, L. Telek and H. D. Graham, eds. AVI Technical Books, Inc., Westport, Connecticut, USA.

Cheeke, P. R., R. Carlsson, and G. O. Kohler. 1981. Nutritive value of leaf protein concentrates prepared from *Amaranthus* species. Canadian Journal of Animal Science 61(1):199–204.

Cole, J. N. 1979. Amaranth from the Past for the Future. Rodale Press, Inc., Emmaus, Pennsylvania, USA. 311 pp., including bibliography of 120 publications.

Coons, M. P. 1982. Relationships of *Amaranthus caudatus*. Economic Botany 36(2):129–146.

Daloz, C. R. 1980. Horticultural aspects of the vegetable amaranthus. M.S. thesis, Cornell University, Ithaca, New York, USA. 150 pp.

Der Marderosian, A. D., J. Beutler, W. Pfender, J. Chambers, R. Yoder, E. Weinsteiger, and J. P. Senft. 1980. Nitrate and oxalate content of vegetable amaranth. Rodale Research Report 80-4. Rodale Press, Inc., Emmaus, Pennsylvania, USA. 15 pp.

Downton, W. J. S. 1973. *Amaranthus edulis*: a high lysine grain amaranth. World Crops 25(1):20.

Edwards, A. D. 1981. Amaranth Grain Production Guide. Rodale Press, Inc., Emmaus, Pennsylvania, USA. 20 pp.

El-Sharkawy, M. A., R. S. Loomis, and W. A. Williams. 1968. Photosynthetic and respiratory exchanges of carbon dioxide by leaves of the grain amaranths. Journal of Applied Ecology 5(1):243–251.

Feine, L. B., R. R. Harwood, C. S. Kauffman, and J. P. Senft. 1979. Amaranth: gentle giant of the past and future. Pp. 41–63 in New Agricultural Crops, G. A. Ritchie, ed. AAAS Selected Symposium 38. Westview Press, Boulder, Colorado, USA.

Foy, C. D. and T. A. Campbell. 1981. Differential tolerances of *Amaranthus* strains to high levels of AL and MN in acid soils. Agronomy Abstracts, 176.

Fuller, H. J. 1949. Photoperiodic responses of *Chenopodium quinoa* willd. and *Amaranthus caudatus* L. American Journal of Botany 36:175–180.

Gilbert, L., and C. S. Kauffman. 1981. Cooking characteristics and sensory qualities of amaranth grain varieties. Rodale Research Report 81-36. Rodale Press, Inc., Emmaus, Pennsylvania, USA. 29 pp.

Grubben, G. J. H. 1976. The cultivation of amaranth as a tropical leaf vegetable, with special reference to South Dahomey. Communication No. 67. Department of Agricultural Research, Royal Tropical Institute, Amsterdam, The Netherlands. 207 pp., including 197 references.

Grubben, G. J. H. 1980. Cultivation methods and growth analysis of vegetable amaranth with special reference to South Benin. Proceedings of the Second Amaranth Conference. Rodale Press, Inc., Emmaus, Pennsylvania. USA.

Grubben, G. J. H., and D. H. van Sloten. 1981. Genetic Resources of Amaranths: A Global Plan of Action. AGP:IBPGR/80/2. International Board for Plant Genetic Resources, Food and Agriculture Organization of the United Nations, Rome, Italy. 57 pp.

Hanelt, P. 1968. Bemerkungen zur Systematik und Anbaugeschichte einiger *Amaranthus*-D Arten. [Contributions to cultivated plants flora. Part 1. Comments on the systematics and the history of cultivation of some Amaranthus-D species.] Die Kulturpflanze 16:127–149.

Harris, D. J., P. R. Cheeke, and N. M. Patton. 1980. A note on the feeding value of *Amaranthus* (pigweed) and *Chenopodium* (lamb's quarters) to rabbits. Journal of Applied Rabbit Research 3(3):11–13.

Haas, P. W. 1983. Amaranth density report. Rodale Research Report NC-83-8. Rodale Press, Inc., Emmaus, Pennsylvania, USA. (A 3-year summary of testing to determine optimum plant populations at Rodale Research Center.)

Hauptli, H., and S. K. Jain. 1980. Genetic polymorphisms and yield components in a population of amaranth. Journal of Heredity 71(4):290–292.

Heiser, C. B., Jr. 1964. Sangorache, an amaranth used ceremonially in Ecuador. American Anthropologist 66:136–140.

Hunziker, A. T. 1952. Los Pseudocereales de la agricultura indígena de América. Acme Agency, Cordoba and Buenos Aires, Argentina. 104 pp.

Irving, D. W., A. A. Betschart, and R. M. Saunders. 1981. Morphological studies on *Amaranthus cruentus*. Journal of Food Science 46:1170–1174.

Jain, S. K., and H. Hauptli. 1980. Grain amaranth: a new crop for California. Agronomy Progress Report No. 107, April 14. Cooperative Extension Service, University of California, Berkeley, California, USA. 3 pp.

Jain, S. K., L. Wu, and K. R. Vaidya. 1980. Levels of morphological and allozyme variation in Indian amaranths, a striking contrast. Journal of Heredity 71(4):283–285.

Joshi, B. D. 1981a. Exploration for amaranth in northwest India. Pp. 41–51 in Plant Genetic Resources Newsletter. AGP:PGR/48. International Board for Plant Genetic Resources, Food and Agriculture Organization of the United Nations, Rome, Italy.

Joshi, B. D. 1981b. Catalogue on amaranth germplasm. Regional Station, National Bureau of Plant Genetic Resources (NBPGR), Phagli, Simla 171012, India. 42 pp.
Joshi, B. D., K. L. Mehra, and S. D. Sharma. 1983. Cultivation of grain amaranth in the northwestern hills (India, *Amaranthus* spp., germplasm, varieties, yields). Indian Farming 32(12):34–35, 37.
Kauffman, C. S. 1982. Improved grain amaranth varieties and their yields. Rodale Research Report NC-81-1. Rodale Press, Inc., Emmaus, Pennsylvania, USA. (A 3-year summary of yield testing at Rodale Research Center.) 45 pp.
Kauffman, C. S., N. N. Bailey, and B. T. Volak. 1983. Amaranth grain production guide. Rodale Research Report NC-83-6. Rodale Press, Inc., Emmaus, Pennsylvania, USA.
Kauffman, C. S., N. N. Bailey, B. T. Volak, L. E. Weber, and N. R. Volk, 1984. Amaranth Grain Production Guide. Rodale Research Report NC-84-6. Rodale Press, Inc., Emmaus, Pennsylvania, USA
Kauffman, C. S., and P. W. Haas. 1982. Grain amaranth: an overview of research and production methods. Rodale Research Report NC-83-5. Rodale Press, Inc., Emmaus, Pennsylvania, USA. 13 pp.
Kauffman, C. S., and C. Reider. 1983. Rodale amaranth germ plasm collection. Rodale Research Report NC-83-2. Rodale Press, Inc., Emmaus, Pennsylvania, USA. 81 pp.
Khoshoo, T. N., and M. Pal. 1972. Cytogenetic patterns in *Amaranthus*. Chromosomes Today 3:259–267.
Koch, B., M. Kota, and I. M. Horvath. 1965. Fodder crops as leaf protein. Agrobotanika 7:19–28.
Leon, J. 1964. Plantas alimenticias Andinos. Pp. 71–85 in Boletin Technico No. 6. Instituto Intermericáno de Ciencias Agrícolas Zona Andina, Lima, Peru.
Lexander, K., R. Carlsson, V. Schalen, A. Simonsson, and T. Lundborg. 1970. Quantities and qualities of leaf protein concentrates from wild species and crop species grown under controlled conditions. Annals of Applied Biology 66(2):193–216.
MacNeish, R. S. 1971. Speculation about how and why food production and village life developed in the Tehuacan Valley, Mexico. Archaeology 24(4):307–315.
Marten, G. C., and R. N. Andersen. 1975. Forage nutritive value and palatability of 12 common annual weeds. Crop Science 15(6):821–827.
Martin, F. W., and L. Telek. 1979. Vegetables for the hot, humid tropics. Part 6. *Amaranthus* and *Celosia*. United States Department of Agriculture, New Orleans, Louisiana, USA. 21 pp., including 15 references.
Mugerwa, J. S., and R. Bwabye. 1974. Yield, composition and in vitro digestibility of *Amaranthus hybridus* subspecies *incurvatus*. Tropical Grasslands 8(1):49–53.
Mugerwa, J. S., and W. Stafford. 1976. Effect of feeding oxalate-rich *Amaranthus* on ovine serum, calcium and oxalate levels. East African Agricultural and Forestry Journal 42(1):71–75.
National Academy of Sciences. 1975. Underexploited Tropical Plants with Promising Economic Value. National Academy of Sciences, Washington, D.C., USA. 189 pp.
Odwongo, W. O., and J. S. Mugerwa. 1980. Performance of calves on diets containing *Amaranthus* leaf meal. Animal Feed Science and Technology 5(3):193–204.
Oliveira de Paiva, W. 1978. Amarantaceas: nova opção de espinafres tropicais para a Amazonia. Acta Amazonica 8:357–363.
Olufolaji, A. O., and T. O. Tayo. 1980. Growth development and mineral contents of three cultivars of amaranth *Amaranthus cruentus*. HortScience 13(2):181–190.
Omueti, O. 1980. Effects of age on celosia cultivars. Experimental Agriculture 16:279–286.
Pal, M., and T. N. Khoshoo. 1974. Grain amaranths. Pp. 129–137 in Evolutionary Studies in World Crops: Diversity and Change in the Indian Subcontinent, Sir J. Hutchinson, ed. Cambridge University Press, London, England.
Pal, M., and T. N. Khoshoo. 1973a. Evolution and improvement of cultivated amaranths. 6. Cytogenetic relationships in grain types. Theoretical and Applied Genetics 43:242–251.
Pal, M., and T. N. Khoshoo. 1973b. Evolution and improvement of cultivated amaranths. 7. Cytogenic relationships in vegetable amaranth. Theoretical and Applied Genetics 43:343–350.

Rajagopal, A., C. R. Muthukrishnan, M. K. Mohideen, and S. Syed. 1977. Co 2 *Amaranthus*. An early vigorous variety. South Indian Horticulture 25(3):102–105.

Rodale Press, Inc. 1977. Proceedings of the First Amaranth Seminar, held at Kutztown, Pennsylvania, USA, July 29, 1977. Rodale Press, Inc., Emmaus, Pennsylvania, USA. 130 pp. (Ten papers; review of research in the United States, 1975–1977, mainly on cereal amaranths.)

Rodale Press, Inc. 1980. Proceedings of the Second Amaranth Conference, held at Kutztown, Pennsylvania, USA, September 13–14, 1979. Rodale Press, Inc., Emmaus, Pennsylvania, USA. 184 pp. (Papers on nutrition, cultivation, assembly, and handling of germplasm.)

Sanchez-Marroquin, A. 1983. Two forgotten crops of agroindustrial importance: amaranth and quinoa. Archivos Latinoamericanos de Nutricion 33:11–32.

Sanchez-Marroquin, A. 1980. Potencialidad agro-industrial del amaranto. Centro de Estudios Economicos y Sociales del Tercer, Mundo, Mexico.

Sauer, J. D. 1967. The grain amaranths and their relatives: a revised taxonomic and geographic survey. Annals of the Missouri Botanical Garden 54(2):103–137.

Sauer, J. D. 1950. The grain amaranths: a survey of their history and classification. Annals of the Missouri Botanical Garden 37:561–619.

Saunders, R. M., and R. Becker. In press. Amaranthus. Vol. 6, Chapter 6 in Advances in Cereal Science and Technology, Y. Pomeranz, ed. American Association of Cereal Chemistry, St. Paul, Minnesota, USA.

Senft, J. P. 1980. Protein quality of amaranth grain. Rodale Research Report 80-3. Rodale Press, Inc., Emmaus, Pennsylvania, USA. 14 pp.

Senft, J. P., C. S. Kauffman, and N. N. Bailey. 1981. The Genus Amaranthus: A Comprehensive Bibliography. Rodale Press, Inc., Emmaus, Pennsylvania, USA. 217 pp.

Singh, H., and A. Rakib. 1971. Chemical evaluation of certain poultry feed stuff. Indian Journal of Animal Research 5(1):39–42.

Singh, H., and T. A. Thomas. 1978. Grain Amaranths, Buckwheat and Chenopods. Indian Council of Agricultural Research, New Delhi, India. 70 pp.

Toll, J., and D. H. van Sloten. 1982. Directory of Germplasm Collections. 4. Vegetables AGP:IBPGR/82/1. International Board for Plant Genetic Resources, Food and Agricultural Organization of the United Nations, Rome, Italy. 187 pp.

Wilson, G. F., and H. P. Curfs. 1976. Cost of production and estimated income from celosia (*Celosia argentea L.*) under two production systems. Vegetables for the Hot, Humid Tropics Newsletter (Mayaguez) 1:35–37.

Appendix B

Research Contacts

Argentina

A. T. Hunziker, Museo Botanico, Facultad de Ciencias Exactas, Fisicas y Naturales, Casilla de Correo 495, 5000 Cordoba

Brazil

L. Jokl and A. Duarte Correa, Faculdade de Farmacia-UFMG, Av. Olegario Maciel 2360, 30000 Belo Horizonte, Minas Gerais (grain protein fractionation, composition)

W. O. dePaiva, Instituto Nacional de Pesquisas da Amazonia (INPA), Caixa Postal 478, Manaus, Amazonas CEP 69.000 (variety testing amaranth, celosia)

Colombia

M. I. Sehvaness, Calle 125 No. 37–47, Bogota

Federal Republic of Germany

F. Mustafa, Roggen Str. 16, 7000 Stuttgart 70 (cultural practices)

German Democratic Republic

P. Hanelt, Zentralinstitut für Genetik und Kulturpflanzenforschung, der Akademie der Wissenschaften der DDR, 4325 Gatersleben

Ghana

J. C. Norman, Head, Department of Horticulture, Dean, Faculty of Agriculture, University of Science and Technology, University P.O., Kumasi (cultivation)

Greece

A. Gagianas, Department of Agronomy, Aristotle University of Thessaloniki, Thessaloniki

Guatemala

R. Bressani, Instituto de Nutricion de Centro America y Panama (INCAP), Apartado Postal 1188, Carretera Roosevelt Zona 11, Guatemala City

India

R. K. Arora, National Bureau of Plant Genetic Resources (NBPGR), Indian Agricultural Research Institute, IARI Campus, New Delhi 110012 (exploration)

B. Choudhury, Division of Vegetable Crops and Floriculture, Indian Agricultural Research Institute (IARI), New Delhi 110012 (variety testing)
R. P. Devadas, Director, Sri Avinashilingam Home Science College for Women, Coimbatore 641043, Tamil Nadu (iron and carotene availability to children)
B. D. Joshi, National Bureau of Plant Genetic Resources (NBPGR), Regional Station, Phagli, Simla 171012 (Indian grain amaranths)
M. Kader Mohideen, Horticultural Research Station, Tamil Nadu Agricultural University, Perumbarai 624212, Madurai District, Tamil Nadu (breeding, cultivation)
T. N. Khoshoo, Secretary to the Government of India, Department of Environment, Bikaner House, Shahjahan Road, New Delhi 110011 (cytogenetics, hybridization)
P. P. Kurien, Central Food Technological Research Institute, Cheluvamba Mansion, V. V. Mohalla, Mysore 570013
C. R. Mathukrishnan [Dean (Horticulture) retired], 13/187 D. B. Road, R. S. Punam, Coimbatore 641002, Tamil Nadu (cultivation, breeding)
G. Oblisami, Department of Agricultural Microbiology, Tamil Nadu Agricultural University, Coimbatore 641003, Tamil Nadu (bacterial inoculants)
M. Pal, Cytogenics Laboratory, National Botanical Research Institute, Lucknow 226001 (cytogenetics, evolution, breeding)
A. Rajagopal, Professor of Agronomy, Agricultural College and Research Institute, Madurai 625104, Tamil Nadu (cultivation)
C. Ramachandran, Department of Olericulture, College of Horticulture, Kerala Agricultural University, Vellanikkara 680654, Trichur, Kerala (horticulture, humid tropic vegetables)
S. Saroja, Sri Avinashilingam Home Science College for Women, Coimbatore 641043, Tamil Nadu (iron and carotene availability to children)
K. G. Shanmugavelu, Professor and Head, Department of Horticulture, Agricultural College and Research Institute, Madurai 625104, Tamil Nadu (breeding)
N. Sivakami, Division of Vegetable Crops and Floriculture, Indian Agricultural Research Institute (IARI), New Delhi 110012 (variety testing)
M. Vijayakumar, Horticultural Officer, Pomological Station, Coonoor, Nilgiris District, Tamil Nadu (varietal testing, growth studies)
M. Vishakantalah, Division of Entomology, University of Agricultural Sciences, Hebbal, Bangalore 560024 (Plutella caterpillars)

Indonesia

Z. Abidin, Balai Penelitian Tanamin Pangan Lembang, Lembang, Jawa Barat (cultivation)
S. Harjadi, Faculty of Agriculture, Department of Agronomy, Bogor Agricultural University, Jalan oto Iskandardinata, Bogor (nutrition, cultivation)

Italy

D. H. van Sloten, Genetic Resources Officer (Horticulture) Crop Genetic Resources Centre/IBPGR Secretariat, Plant Production and Protection Division, FAO, Via delle Terme di Caracalla, 00100 Rome (genetic resources)

Kenya

V. K. Gupta, Department of Crop Science, University of Nairobi, P.O. Box 30197, Nairobi
S. K. Imbamba, Department of Botany, University of Nairobi, P.O. Box 30197, Nairobi (protein analysis)
J. O. Kokwaro, Department of Botany, University of Nairobi, P.O. Box 30197, Nairobi (taxonomy and ecology)

Mexico

A. Sanchez-Marroquin, Instituto Nacional de Investigaciones Agricolas, Miami 40, Mexico, D.F. 03810 (industrial uses of amaranth grain)

A. Trinidad-Santos, Centro de Edafologia, Colegio de Postgraduados, Chapingo, Edo. de Mexico (agronomic research)

Nepal

H. K. Saiju, Royal Botanical Gardens, Godawary, Lalit pur Dist. (cereal amaranth germplasm)

Netherlands

G. J. H. Grubben, Research Station for Arable Farming and Field Production of Vegetables, Edelhertweg 1, Postbus 430, 8200 AK Lelystad

New Zealand

R. A. Crowder, Lincoln College, Canterbury
D. W. Devine, Product Development Manager, N.Z. Flourmills, Ltd., P.O. Box 30461, Lower Hutt

Nigeria

O. Bassir, Biochemistry Department, University of Ibadan, P.O. Box 4021, University Post Office, Oyo Road, Ibadan (protein)
L. Denton, National Horticultural Research Institute, Idi-Ishin, P.M.B. 5432, Ibadan (germplasm collection, breeding)
A. A. O. Edema, Breeder/Geneticist, National Horticultural Research Institute, P.M.B. 5432, Ibadan (amaranth and celosia cultivation)
M. Fafunso, Department Biochemistry, University of Ibadan, Ibadan (food technology, composition, vitamin C, amaranth, celosia)
International Institute for Tropical Agriculture (IITA), Oyo Road, P.M.B. 5320, Ibadan (agronomy)
J. O. S. Kogbe, Institute of Agricultural Research and Training, University of Ife, P.M.B. 5029, Moor Plantation, Ibadan
O. L. Oke, Dean, Faculty of Science, University of Ife, Ile-Ife (nutritional value)
B. N. Okigbo, Deputy Director General, International Institute for Tropical Agriculture (IITA), Oyo Road, P.M.B. 5320, Ibadan (farming systems)
A. O. Olufolaji, National Horticultural Research Institute, P.M.B. 5432, Ibadan (cultivation, mineral content)
O. Omueti, Institute of Agricultural Research and Training, University of Ife, P.M.B. 5029, Moor Plantation, Ibadan (celosia cultivation)
I. C. Onwueme, School of Agricultural Technology, Federal University of Technology, P.M.B. 1526, Owerri (temperature stress, amaranth and celosia)
O. Osi Banjo, Department of Chemistry, University of Ibadan, Ibadan (vitamin C)
Prem Nath, National Horticultural Research Institute (NIHORT), Idi-Ishin, P.M.B. 5432, Ibadan (germplasm)
T. O. Tayo, Department of Agricultural Biology, University of Ibadan, Ibadan (cultivation, mineral content)
G. F. Wilson, International Institute for Tropical Agriculture (IITA), Oyo Road, P.M.B. 5320, Ibadan (cultivation)

Peru

S. E. Antunez de Mayolo R., Apartado (P.O.B.) 18-5469, Lima 18
R. Ferreyra, Museo de Historia Natural, Universidad Nacional Mayor de San Marcos, Casilla 11434, Lima 14
L.S. Kalinowski, Department of Agriculture, Universidad Nacional de Cuzco, Avenida Infancia 440, Cuzco

Sierra Leone

Sama S. Monde, Biological Sciences Department, Njala University College, P.M.B., Freetown (grain)

Sweden

R. Carlsson, Department of Plant Physiology, Box 7007, S-220 07, Lund (whole plant, leaf, leaf nutrient/protein concentrate, grain: composition, nutritive value)

Taiwan

Han Huang, Department of Horticulture, National Taiwan University, Taipei 107 (temperature)
L. Ho, AVRDC Seed Laboratory, P.O. Box 42, Shanhua, Tainan (germplasm collection)
Lin Chao-Hsiung, Department of Vegetable Crops, Fengshan Tropical Horticultural Experiment Station, Fengshan, Kaohsiung (cultivation, variety testing)

Tanzania

N.A. Mnzava, Faculty of Agriculture, Forestry and Veterinary Sciences, University of Dar es Salaam, Chuo Kikuu, Morogoro

Thailand

Chuckree Senthong, Chairman, Department of Agronomy, Chiang Mai University, Chiang Mai
Soonthorn Duriyaprapan, Thailand Institute of Scientific and Technological Research (TISTR), 196 Phahonyothin Road, Bangkhen, Bangkok 9
T. Tonguthaisri, Mae Jo Institute of Agricultural Technology, Chiang Mai (vegetable amaranths collection)

United Kingdom

G. J. L. Griffin, Ecological Materials Research Institute, Brunel University, Shoreditch Campus, Egham, Surrey TW20 OJZ
B. Pickersgill, Department of Agricultural Botany, Plant Science Laboratories, University of Reading, Whiteknights, Reading RG6 2AS
C. C. Townsend, Herbarium, Royal Botanic Gardens, Kew, Richmond, Surrey
L. St. Lawrence, Kins Plants Ltd., Woodcote Grove Ashley Road, Epsom, Surrey KT18 5BW

United States

G. C. W. Ames, Department of Agricultural Economics, University of Georgia College of Agriculture, 301 Conner Hall, Athens, Georgia 30602 (Vegetable amaranth in Zaire)
W. Applegate, Post Rock Natural Grain, Box 24A, Luray, Kansas 67649 (farm research and production of grain amaranth)
N. N. Bailey, Rodale Research Center, R.D. 1, Box 323, Kutztown, Pennsylvania 19530
A. R. Baldwin (Retired, Vice President and Executive Director of Research, Cargill, Inc.), 4854 Thomas Avenue South, Minneapolis, Minnesota 55410
R. Becker, USDA Western Regional Research Laboratory, Cereals Research Unit, 800 Buchanan St., Berkeley, California 94710 (nutrition, composition)
J. A. Beutler, School of Pharmacy, Auburn University, Auburn, Alabama 36849 (feeding trials, nitrate and oxalate)
A. Betschart, Nutrients Research Unit, USDA Western Regional Research Laboratory 800 Buchanan St., Berkeley, California 94710
T. A. Campbell, Germplasm Resources Laboratory, Beltsville Agricultural Research Center, Beltsville, Maryland 20705 (plant introduction and vegetable amaranth research)

M. P. Coons, 6715 SW 88th Street, No. 712, Miami, Florida 33156 (taxonomy)
P. R. Cheeke, Department of Animal Science, Oregon State University, Corvallis, Oregon 97331 (feeding trials, grain and leaves)
C. R. Daloz, R.D. 1, Box 819, Hancock, New Hampshire 03449
A. H. DerMarderosian, Philadelphia College of Pharmacy and Science, Philadelphia, Pennsylvania 19104 (feeding trials)
D. K. Early, Central Oregon Community College, N.W. College Way, Bend, Oregon 97701
A. D. Edwards, Agronomy Department, Virginia Polytechnic Institute, Blacksburg, Virginia 24060
J. Ehleringer, Department of Biology, 201 Biology Building, University of Utah, Salt Lake City, Utah 84112
L. B. Feine, Rodale Research Center, R. D. 1, Box 323, Kutztown, Pennsylvania 19530 (germplasm, taxonomy)
H. Flores, Department of Biology, Yale University, New Haven, Connecticut 06520 (physiology, tissue culture)
A. W. Galston, Department of Biology, Yale University, New Haven, Connecticut 06520 (physiology, tissue culture)
L. Gilbert, Test Kitchen, Rodale Press, 33 E. Minor Street, Emmaus, Pennsylvania 18049 (amaranth foods)
R. R. Harwood, Rodale Research Center, R.D. 1, Box 323, Kutztown, Pennsylvania 19530 (cultivation practices, breeding)
P. W. Haas, Rodale Research Center, R.D. 1, Box 323, Kutztown, Pennsylvania 19530 (cultivation practices, gremplasm)
H. Hauptli, Agronomy and Range Science Department, University of California, Davis, California 95616 (genetics, germplasm exploration Central and South America)
M. Irwin, 2823 Oakridge Avenue, Madison, Wisconsin 53704 (production of grain amaranth)
S. K. Jain, Agronomy and Range Science Department, University of California, Davis, California 95616 (genetics, evolution)
M. Jones, Rt. 1, Lodgepole, Nebraska 69419 (production of grain amaranth)
C. S. Kauffman, Rodale Research Center, R.D. 1, Box 323, Kutztown, Pennsylvania 19530 (breeding cereal amaranths, germplasm)
M. Langley, P.O. Box 9085, Austin, Texas 78766
R. D. Locy, NPI, Inc., 417 Wakara Way, Salt Lake City, Utah 84108
J. Martineau, Plant Resources Institute, 360 Wakara Way, Salt Lake City, Utah 84108
J. B. McElroy, Department of Plant Breeding and Biometry, Cornell University, Ithaca, New York 14853 (breeding taxonomy)
C. M. McKell, Vice President, Research, NPI, Inc., 417 Wakara Way, Salt Lake City Utah 84108
C. McNeil, RR 1, Box 30, Paradise, Kansas 67658 (production of grain amaranth)
W. P. Miller, Director, Amerind Agrotech Laboratory, P.O. Box 97, Sacaton, Arizona 85247
H. M. Munger, 410 Bradfield Hall, Cornell University, Ithaca, New York 14853 (breeding)
G. Nabhan, President, Native Seeds—Search, 3950 West New York Drive, Tucson, Arizona 85745 (germplasm, ethnobotany)
T. Ney, Rodale Food Center, Rodale Press, 33 E. Minor Street, Emmaus, Pennsylvania 18049 (amaranth foods)
Pai Chi Chang, Rt. 30, Williamstown, New York 13493
K. Patchen, RR 2, Box 396A, Mundelain, Illinois 60060 (production of grain amaranth)
M. L. Price, ECHO, R.R. 2, Box 852, North Fort Myers, Florida 33903
R. Ramback, Applegate Produce, 3030 Upper Applegate, Jacksonville, Oregon 97530 (production of grain amaranth)
J. F. Ramirez, New York State Agriculture Experiment Station, P.O. Box 462, Food Research Center, Geneva, New York 14456
C. Reider, Rodale Research Center, R. D. 1, Box 323, Kutztown, Pennsylvania 19530 (germplasm, yield studies)
J. R. K. Robson, 171 Ashley Ave., Charleston, South Carolina 29403
R. Rodale, Rodale Press, 33 E. Minor Street, Emmaus, Pennsylvania 18049

R. M. Ruberte, Mayagüez Institute of Tropical Agriculture, USDA, Box 70, Mayagüez, Puerto Rico 00708 (botany, composition)

D. J. Sammons, Department of Agronomy, University of Maryland, College Park, Maryland 20742

R. M. Saunders, Cereal Products Research, USDA Western Regional Research Laboratory, 800 Buchanan St., Berkeley, California 94710

J. Senft, 378 Fairview St., Emmaus, Pennsylvania 18049 (nitrate and oxalate studies, protein quality)

M. Shannon, U.S. Salinity Laboratory, USDA-ARS, 4500 Glenwood Drive, Riverside, California 92501

A. A. Sigle, R.R. 1, Box 2, Luray, Kansas 67649 (farm research and production of grain amaranth)

L. Telek, Mayagüez Institute of Tropical Agriculture, USDA, Box 70, Mayagüez, Puerto Rico 00708 (botany, composition)

K. R. Vaidya, Department of Agronomy and Range Science, University of California, Davis, California 95616 (genetics)

D. Wall, Department of Agronomy, University of Maryland, College Park, Maryland 20742

L. Walters, 1640 Harris Lane, Naperville, Illinois 60565 (production of grain amaranth)

L. Wu, Environmental Horticulture Department, University of California, Davis, California 95616 (genetics)

Zambia

Zambia Seed Co, Ltd., P.O. Box 35441, Buyantanshi Road, Lusaka (cultivation, germplasm)

Appendix C

Germplasm Collections and Commercial Seed Suppliers

The following list of amaranth seed sources is based on Grubben and van Sloten, 1981, and Toll and van Sloten, 1982. See Selected Readings.

Benin

Centre de Formation Horticole et Nutritionnelle, B.P. 13, Ouando, Port-Novo (commercial seed; both amaranth and celosia cultivars)

Bolivia

Estacion Experimental de Patacamaya, Instituto Boliviano de Tecnologia Agropecuaria (IBTA), Patacamaya (N. Lizarraga, *Amaranthus caudatus*, 430 landraces from the Andean region)

France

Institut National de Recherches Agronomiques, Petit-Bourg, Guadeloupe (working collection of local cultivars)

German Democratic Republic

Zentralinstitut für Genetik und Kulturpflanzenforschung, Corrensstrasse 3, 4325 Gatersleben (H. Böhme, Chr. Lehmann, 100 samples of 17 wild, weedy, and cultivated species)

Ghana

Department of Horticulture, University of Science and Technology, Kumasi (working collection)

Hong Kong

Wong Yukhop Seed Co., 20 Pei Ho Street, Shumshuipo, Kowloon (two commercial cultivars)

India

Pocha's Seeds, P.O. Box 55, Poona 411001, Maharashtra (3 commercial cultivars)
Division of Vegetable Crops and Floriculture, Indian Agricultural Research Institute (IARI), New Delhi 110012 (working collection)

National Bureau of Plant Genetic Resources (NBPGR), Indian Agricultural Research Institute, IARI Campus, New Delhi 110012 (K. L. Mehra; *Amaranthus* species, 824 mostly landraces from India)
Division of Vegetable Crops, Institute of Horticultural Research, 225 Upper Palace Orchards, Bangalore (working collection)
Kerala Agricultural University, College of Horticulture, P.O. Vellanikkara 680651, Trichur Kerala (working collection)
Faculty of Horticulture, Tamil Nadu Agricultural University, Coimbatore 641003 (C. P. Muthukrishnan; 450 landraces mainly from India)
National Botanical Gardens, Lucknow, 226001, UP (working collection)
National Bureau of Plant Genetic Resources (NBPGR), Regional Station, Phagli, Simla 171012 (B. D. Joshi; more than 1,000 samples of landraces from India and 27 introductions)
Suttons Seeds, P.O. Box 9010, Calcutta-16 (4 commercial cultivars)

Indonesia

National Biological Institute, P.O. Box 110, Bogor (S. Sastrapradja; 75 landraces from Indonesia)

Japan

T. Sakata, CPO Box Yokohama 11, Yokohama 220-91 (1 commercial cultivar)

Netherlands

Department of Tropical Crops, Agricultural University, Ritzema Bosweg 32, Wageningen (working collection)

Nigeria

International Institute for Tropical Agriculture (IITA), Oyo Road, P.M.B. 5320, Ibadan (working collection)
Plant Science Department, University of Ife, Ile-Ife (working collection)
National Horticultural Research Institute (NIHORT), Idi-Ishin, P.M.B. 5432 Ibadan (T. Badra, A. A. O. Edema, L. Denton; 240 population samples of landraces and breeders' lines from Nigeria)

Peru

Estacion Experimental de Camacani, Universidad Nacional Tecnica del Altiplano, Puno (L. Lescano, L. Perez; 440 landraces from the Andean region)
Centro de Investigacion de Cultivos Andinos, Universidad Nacional del Cuzco, Avenida de la Infancia 440, Huanchac, Cuzco (L. Sumar Kalinowski; 100 accessions cereals)

Taiwan

Asian Vegetable Research and Development Center (AVRDC), P.O. Box 42, Shanhua, Tainan 741 (L. Ho; 92 samples of wild forms, landraces, and commercial cultivars from Africa, Asia, and the United States)
Hsing Nong Seed Co., 188 Sec. 4 Chung Hsin Road, P.O. Box San Chung No. 2, San Chung, Taipei (several commercial cultivars)
Known-You Seed Co., 26 Chung Cheng 2nd Road, Ka Ohsiung, Taiwan (popular cultivars for Taiwan)
Taipei District Agricultural Improvement Station, Chin Chun, Taipei (working collection)
Taiwan Seed Service, Shin-Shien, Taichung (several commercial cultivars)

Thailand

Chia Tai Seeds and Agricultural Company, Ltd., 295-303 Songsawad Road, Bangkok (2 commercial cultivars)

Fang Horticultural Experiment Station, Department of Agriculture, Fang, Chiang Mai (T. Thonguthaisri; 115 landraces from Thailand)

United Kingdom

Royal Botanic Gardens, Kew, Wakehurst Place, Ardingly, Haywards Heath, Sussex RH17 6TN (S. Linington; samples of wild origin)

United States

Burpee Seed Company, Warminster, Pennsylvania (1 commercial cultivar; tampala)

G. Seed Co., P.O. Box 702, Tonasket, Washington 98855

Germplasm Resources Laboratory, Beltsville Agricultural Research Center, Beltsville, Maryland 20705 (gene bank, plant introduction for long-term storage)

Grace's Gardens, 22 Autumn Lane, Hackettstown, New Jersey 07840

Gurney Seed and Nursery Co., Yankton, South Dakota 57079

Johnny's Selected Seeds, Albion, Maine 04910

Mellinger's Inc., 2310 W. So. Range Road, North Lima, Ohio 44452

National Seed Storage Laboratory (NSSL), U.S. Department of Agriculture, Colorado State University, Fort Collins, Colorado 80523 (L. N. Bass; duplicates of Rodale collection)

Park Seed Company, P.O. Box 31, Greenwood, South Carolina 29647, (1 commercial cultivar; Tampala)

Plants of the Southwest, 1570 Pachero St., Santa Fe, New Mexico 87501

Redwood City Seed Co., P.O. Box 361, Redwood City, California 94064

Rodale Research Center, R.D. 1, Box 323, Kutztown, Pennsylvania 19530 (C. S. Kauffman; more than 600 accessions cereals, vegetable, and wild species)

Thompson and Morgan, P.O. Box 100, Farmingdale, New York 07727 (1 commercial cultivar; Hinn Choy)

Tsang and Ma International, P.O. Box 294, Belmont, California 94002 (1 commercial cultivar)

Zambia

Crop Science Department, School of Agricultural Sciences, University of Zambia, P.O. Box 2379, Lusaka (collection of 100 local cultivars plus landraces from India, Nigeria, and the United States)

Appendix D

Biographical Sketches of Panel Members

MELVIN G. BLASE, Professor of Agricultural Economics, University of Missouri, Columbia, received his Ph.D. from Iowa State University in 1960 and served on the faculties of that institution and the Air Force Institute of Technology before joining the University of Missouri. While he has undertaken research in international agricultural development in Asia, Africa, and Latin America, his domestic research centers on new crops for U.S. agriculture and alternative energy sources. Dr. Blase has written one book, edited two others, and written numerous articles.

T. AUSTIN CAMPBELL is a Research Agronomist with the U.S. Department of Agriculture and is currently specializing in new-crop breeding. He received his B.S. in conservation and resource development from the University of Maryland in 1967 and spent six years with the alfalfa project of the USDA as a technician and research assistant. He received an M.S. in plant breeding from the University of Maryland in 1972 and joined the Weed Science Laboratory as a support scientist in 1973. Two years later he assumed his present position conducting research on numerous potential new-crop species, including kenaf, *Crambe, Lesquerella, Limnanthes, Asclepias syriaca, Rhus glabra, Cuphea,* and Stokes aster, as well as amaranth. He received his Ph.D. in plant breeding from the University of Maryland in 1980.

LAURIE B. FEINE, an amaranth taxonomy and germplasm consultant, earned her B.A. from the University of Colorado in environmental, population, and organismic biology in 1976. She worked for the Rodale Research Center, specializing in amaranth germplasm and taxonomy, and began plant-breeding and selection work for improved grain varieties. She continued her work on amaranth taxonomy at Harvard Herbaria and has also collected amaranth germplasm in Mexico and Peru for the Rodale Research Center and IBPGR.

HECTOR E. FLORES-MERINO, Research Assistant at the Department of Biology, Yale University, received a B.S. in biology in 1974 from the Universidad Nacional Mayor de San Marcos in Lima, Peru. After teaching for two years, he did research at the Puerto Rico Nuclear Center in Mayagüez and earned an M.S. in horticulture from the College of Agriculture at Mayagüez in 1978. He was a research assistant at Rutgers University, Department of Horticulture, and is currently completing his doctoral studies at Yale. His research has involved propagation of ornamental and crop plants through tissue and cell culture and development of inoculants for nitrogen-fixing bacteria. He is currently studying the physiology and biochemistry of polyamines in higher plants and metabolic responses of plant cells to various types of environmental stress.

LINDA C. GILBERT, Coordinating Supervisor of Product Development in the Test Kitchen at Rodale Press, Inc., received a B.Ag.Sci. degree from the University of Arizona in 1978. Since that time, she has worked in the Rodale Test Kitchens developing healthful recipes for magazines and books. Her research has centered primarily around uses for, and varietal selections of, vegetable protein food sources, including grain and vegetable amaranths, vegetable soybeans, sprouting soybeans, okra seed, and cold-tolerant leafy green vegetables.

GERALD J. H. GRUBBEN is head of the Department of Crop Research at the Research Station for Arable Farming and Field Production of Vegetables in Lelystad, Netherlands. He served as a FAO expert in horticultural projects in Ivory Coast and Benin from 1965 to 1973. In 1975 he received his Ph.D. (agriculture) on a thesis "The Cultivation of Amaranth as a Tropical Leaf Vegetable." From 1975 to 1981 he was attached to the Department of Agricultural Research, Royal Tropical Institute, Amsterdam. He conducted a survey on genetic resources of vegetables for the International Board of Genetic Resources. He was involved in consultancy missions on vegetable growing, crop research, germplasm, and seed production to tropical countries.

RICHARD R. HARWOOD, Director of the Rodale Research Center in Kutztown, Pennsylvania, received a B.S. from Cornell University and a Ph.D. in horticulture and plant breeding from Michigan State University in 1967. He worked as a staff member of the Rockefeller Foundation in India and Thailand from 1967 to 1971 and then as head of the Cropping Systems Program at the International Rice Research Institute until 1976. Dr. Harwood has had numerous Third World agricultural development consulting assignments. He participated as a cropping system specialist in the 1976 vegetable cropping systems delegation of the Committee on Scholarly Exchange with the People's Republic of China. His current work is focused on organic farming systems and new crop development for a sustainable agriculture.

SUBODH JAIN, Professor of Agronomy at the University of California, Davis, received a B.S. from Delhi University in 1954 and a Ph.D. from the University of California, Davis, in 1960. He has worked extensively on the population biology of inbreeding crop species from which he has recently developed interests in the evaluation and conservation of genetic resources and in the domestication of new crops. He has traveled widely for both germplasm collections and teaching/consulting assignments and has had Guggenheim and Fulbright awards for work in Australia and India, respectively.

CHARLES S. KAUFFMAN, Coordinator of New Crops Research at the Rodale Research Center in Kutztown, Pennsylvania, received his B.S. in horitculture from the Pennsylvania State University in 1971 and an M.S. in horticulture and plant breeding from North Carolina State University in 1974. He worked in an agricultural development project for unconventional crops in the southeastern United States for Thomas J. Lipton, Inc. Since 1978, the majority of his time has been spent doing research and writing on grain amaranth, especially as related to germplasm cataloging, varietal development, and the improvement of cultural techniques.

T. N. KHOSHOO is Secretary to the Government of India, Department of Environment. He received his B.Sc. and M.Sc. (botany) from Panjab University, Lahore, and his Ph.D. (cytogenetics of conifers) from Chandigarh. He has served as Director of the National Botanical Research Institute and Deputy Director of the National Botanic Gardens in Lucknow. He worked on the experimental evolution and improvement of nonagricultural economic plants, particularly ornamental and subsidiary food plants (amaranths). He also worked out the genetic-evolutionary race histories and evolved several new cultivars that have sold in nursery trade. He is now concerned with the policy, planning, and management of the environment in India, including wildlife.

JUDITH M. LYMAN, College of Agriculture and Life Sciences, Cornell University, Ithaca, New York, received her Ph.D. in plant breeding in 1980, her M.S. in floriculture and horticulture in 1976, and her B.A. in botany at Duke University, Durham, North Carolina. After two years of field work at the Centro International de Agricultura Tropical (CIAT) in Cali, Colombia, as Visiting Research Associate and Assistant to the Director of Research, she is currently Visiting Research Fellow at the Rockefeller Foundation, where she is involved in international agricultural program activities and project germplasm resources from biological and economic perspectives.

CYRUS M. MCKELL is Vice-President, Research, Native Plants, Inc., Salt Lake City, Utah. He received his B.S. and M.S. degrees from the University of Utah in biological science and botany. In 1956 he earned a Ph.D. in plant ecology, with minors in soils and rangeland management from Oregon State University. From 1956 to 1961 he served as a range plant physiologist with the USDA Agricultural Research Service at Davis, California. In 1961 he became Vice-Chairman and later Chairman of the Agronomy Department, University of California, Riverside, and conducted research on arid land management problems. Subsequently, he joined the Range Science Department at Utah State University as department head, and later as Director of the Institute for Land Reclamation, a post he held until 1980. He has held numerous consultancies on arid land management worldwide and was a Fulbright scholar to Spain in 1968.

GARY PAUL NABHAN is the President of Native Seeds/SEARCH in Tucson, Arizona. He received his Ph.D. from the University of Arizona in 1983 and is a principal investigator conducting research on native agricultural ecosystems through the Office of Arid Lands Studies, University of Arizona. While receiving his M.S. in plant sciences at the same university, he assembled important germplasm collections of tepary beans, sunflowers, devil's claw, and other desert-adapted crops. He has also collected amaranths from the Sierra Madre and the Valley Mexico, as well as from U.S. Indian reservations. He has published more than twenty articles and one book *(The Desert Smells Like Rain)* on desert foods, ethnobotany, and seed conservation. Dr. Nabhan is on the founding board of directors of the Society of Ethnobiology and its journal and is honorary Vice-President of the Seed Saver's Exchange.

DONALD L. PLUCKNETT, with the Consultative Group on International Agricultural Research, the World Bank, received B.S. and M.S. degrees in agriculture and agronomy from the University of Nebraska in 1953 and 1957, respectively, and a Ph.D. in tropical soil science from the University of Hawaii in 1961. He has worked extensively in tropical crop and pasture research and has had broad international experience in tropical agriculture. He has been a consultant for many international groups, including working for the Ford Foundation on the Aswan Project in Egypt, for the United Nations Food and Agriculture Organization, Consultative Group on International Agricultural Research, United States Agency for International Development (USAID), and the South Pacific Commission. From 1973 to 1976 he was Chief of the Soil and Water Management Division, Office of Agriculture, Technical Assistance Bureau, Agency for International Development, Washington, D.C. In 1976 he was awarded AID's Superior Honor Award for his activities in International Development. He has served on several National Academy of Sciences' study panels.

HUGH POPENOE is Professor of Soils, Agronomy, Botany, and Geography and Director of the Center for Tropical Agriculture and International Programs (Agriculture) at the University of Florida. He received his B.S. from the University of California, Davis, in 1951 and his Ph.D. in soils from the University of Florida in 1960. His principal research interest has been in the area of tropical agriculture and land use. Dr. Popenoe's early work in shifting cultivation is one of the few contributions to knowledge of this system. He has traveled and worked in most of the countries in the tropical areas of Latin America, Asia, and Africa. He has served as Director of the Florida Sea Grant College, is a member of the Board of Trustees of the Escuela Agricola Panamericana in Honduras, and was Chairman of the Board for several years. He is a Visiting Lecturer on Tropical Health at the Harvard School of Public Health and is a Fellow of the American Association for the Advancement of Science, the American Society of Agronomy, the American Geographical Society, and the International Soils Science Society. He has served as Chairman of the Joint Research Committee of the Board for International Food and Agricultural Development (Title XII). He was Chairman of the Advisory Committee for Technology Innovation and a member of the Board on Science and Technology for International Development of the National Academy of Sciences.

ALFREDO SANCHEZ-MORROQUIN is a Research Scientist with the CIAMEC division of INIA (National Institute of Agriculture Research) in Chapingo, Mexico, and technical adviser of four agroindustries of Mexico and South America. After completing his B.Sc. in chemistry and microbiology at the National Polytechnic Institute, School of Biological Sciences, and his M.S. from Northwestern University, he joined the Polytechnic Institute as Professor of Criptogamic Botany and later as Chairman of the Department of Microbiology. In 1951 he received a D.Sc. degree from the National University of Mexico when he was working at the Faculty of Chemistry as Professor of Chemical Microbiology and later as Chairman of the Biology Department. He also worked with the School of Agronomy and the Post-Graduate College of the Secretary of Agriculture and was invited to be a visiting professor and scientific investigator in several South American universities and technological institutes. His research has focused on biotechnology: single-cell protein, fermentations, plant products, and food processing. He has published more than 100 papers, five books, and several journal articles and booklets. In 1979 he was appointed Emeritus Professor by the National Polytechnic Institute. In the same year he was awarded three national prizes in science and technology.

JONATHAN D. SAUER is Professor of Geography at the University of California, Los Angeles. He received a B.A. in 1939 from the

University of California, Berkeley, and his Ph.D. in botany at Washington University, St. Louis, and the Missouri Botanical Garden in 1950. From then until 1967, when he joined UCLA, he was on the botany faculty of the University of Wisconsin, Madison. His dissertation was on the grain amaranths, and he has continued to be interested in them and their wild relatives, trying to identify specimens sent in for taxonomic determination. His current research is primarily on ecology and geography of tropical seashore vegetation.

ROBIN M. SAUNDERS is Research Leader of the Cereals Research Unit, Western Regional Research Center, USDA, Albany, California. He has a B.S. and Ph.D. in chemistry from Birmingham (1960) and Newcastle (1963) universities, England, respectively. In 1966, after two years of postdoctoral work in biochemistry at the University of California, Berkeley, and one year in medical research in Pennsylvania, he joined USDA. In his current position, he directs a large research group engaged in various aspects of cereal-grain utilization, including development of crops tolerant to drought, temperature, and saline stress.

JOSEPH P. SENFT, formerly coordinator of nutritional programs at the Rodale Research Center, received a B.S. degree in biology from Juniata College in 1959 and M.S. and Ph.D degrees in biology from the State University of New York, Buffalo, in 1961 and 1965, respectively. His research has been on the evaluation of nutritional quality of both grain and vegetable amaranth. His research in agriculture has focused on soil–plant nutritional relationships.

ARRIS A. SIGLE is a farmer in north central Kansas. He received his B.S. degree in agricultural engineering from Kansas State University in 1973. He has been working with grain amaranth test plots and research since 1978. In 1981 he harvested and successfully marketed 14 acres of amaranth. He also grows wheat and milo and is in charge of a 200-head flock of ewes. He is especially interested in repairing, modifying, and designing new equipment to make his work with these projects more efficient.

THEODORE W. SUDIA is Senior Scientist, National Park Service. He received his B.S. from Kent State University in 1950 and his M.S. and Ph.D. from Ohio State University in 1951 and 1954, respectively. He taught at Winona State College, Winona, Minnesota, and at the University of Minnesota where he was in the Department of Plant Pathology and Physiology. From 1967 to 1969 he was the Associate Director of the American Institute of Biological Sciences. In 1969 Dr. Sudia joined the National Park Service as Research Biologist. He has served as Chief, Ecological Services Division; Chief Scientist and Associate Director for Science and Technology of the Park Service; and Deputy Science Advisor of the National Park Service.

His research interests have ranged from plant ecology to environmental physiology. He is a Fellow of the AAAS.

JAMES L. VETTER is Vice-President, Technical, of the American Institute of Baking, a nonprofit research and educational organization. Dr. Vetter received an A.B. degree with a major in chemistry from Washington University in St. Louis in 1954 and his M.S. and Ph.D. in food technology from the University of Illinois in 1955 and 1958, respectively. Before joining the American Institute of Baking in 1977, he had a 10-year industrial research career working for companies in or related to the baking industry. These companies include Monsanto Company, Standard Brands, and Keebler Company. Dr. Vetter's current responsibilities involve administration of research activities related to the nutrition and science and technology of baking.

DAVID ERVIN WALSH is the Director and Vice-President of Research of the General Nutrition Corporation. He received his B.A. from St. Cloud State University in 1961 and an M.A. and Ph.D. from North Dakota State University, where he was an Associate Professor of Cereal Chemistry and Technology until 1974, when he joined the staff of General Nutrition Corporation in Fargo, North Dakota. His work includes computerization of food processing, research on lipids of barley, proteins of wheat, and on the industrial utilization of wheat. His present research includes: directing the corporation research program in food and cosmetic development, nutrition research on food supplements and health, and coordination of grants and aid programs for academic research directed toward food supplements and nutrition.

NOEL D. VIETMEYER, staff officer for this study, is Professional Associate of the Board on Science and Technology for International Development. A New Zealander with a Ph.D. in organic chemistry from the University of California, Berkeley, he now works on innovations in science that are important for developing countries.

Advisory Committee on Technology Innovation

HUGH POPENOE, Director, International Programs in Agriculture, University of Florida, Gainesville, Florida (Chairman through 1983)

ELMER L. GADEN, JR., Department of Chemical Engineering, University of Virginia, Charlottesville, Virginia, *Chairman*

Members

WILLIAM BRADLEY, Consultant, New Hope, Pennsylvania (through 1983)

CARL N. HODGES, Director, Environmental Research Laboratory, Tucson, Arizona

RAYMOND C. LOEHR, Director, Environmental Studies Program, Cornell University, Ithaca, New York

CYRUS M. MCKELL, NPI, Inc., Salt Lake City, Utah

DONALD L. PLUCKNETT, Consultative Group on International Agricultural Research, Washington, D.C.

EUGENE B. SHULTZ, JR., Professor of Engineering and Applied Science, Washington University, St. Louis, Missouri

THEODORE SUDIA, Deputy Science Advisor to the Secretary of the Interior, Department of the Interior, Washington, D.C. (through 1983)

Board on Science and Technology for International Development

RALPH HERBERT SMUCKLER, Dean of International Studies and Programs, Michigan State University, East Lansing, Michigan, *Chairman*

Members

SAMUEL P. ASPER, President, Educational Commission for Foreign Medical Graduates, Washington, D.C.

DAVID BELL, Department of Population Sciences, Harvard School of Public Health, Boston, Massachusetts

LAWRENCE L. BOGER, President, Oklahoma State University, Stillwater, Oklahoma

ROBERT H. BURRIS, Department of Biochemistry, University of Wisconsin, Madison, Wisconsin

CLAUDIA JEAN CARR, Conservation and Resource Studies, University of California at Berkeley, Berkeley, California

NATE FIELDS, Director, Developing Markets, Control Data Corporation, Edina, Minnesota

ROLAND J. FUCHS, Chairman, Department of Geography, University of Hawaii at Manoa, Honolulu, Hawaii, *ex officio*

ELMER L. GADEN, JR., Department of Chemical Engineering, University of Virginia, Charlottsville, Virginia

The National Academy of Sciences

The National Academy of Sciences was established in 1863 by Act of Congress as a private, nonprofit, self-governing membership corporation for the furtherance of science and technology, required to advise the federal government upon request within its fields of competence. Under its corporate charter the Academy established the National Research Council in 1916, the National Academy of Engineering in 1964, and the Institute of Medicine in 1970.

The National Research Council

The National Research Council was established by the National Academy of Sciences in 1916 to associate the broad community of science and technology with the Academy's purposes of furthering knowledge and of advising the federal government. The Council operates in accordance with general policies determined by the Academy under the authority of its congressional charter of 1863, which establishes the Academy as a private, nonprofit, self-governing membership corporation. The Council has become the principal operating agency of both the National Academy of Sciences and the National Academy of Engineering in the conduct of their services to the government, the public, and the scientific and engineering communities. It is administered jointly by both Academies and the Institute of Medicine. The National Academy of Engineering and the Institute of Medicine were established in 1964 and 1970, respectively, under the charter of the National Academy of Sciences.

The Office of International Affairs

The Office of International Affairs is responsible for many of the international activities of the Academy and the Research Council. Its primary objectives are to enhance U.S. scientific cooperation with other countries; to mobilize the U.S. scientific community for technical assistance to developing nations; and to coordinate international projects throughout the institution.

The Board on Science and Technology for International Development

The Board on Science and Technology for International Development (BOSTID) of the Office of International Affairs addresses a range of issues arising from the ways in which science and technology in developing countries can stimulate and complement the complex processes of social and economic development. It oversees a broad program of bilateral workshops with scientific organizations in developing countries and conducts special studies. BOSTID's Advisory Committee on Technology Innovation publishes topical reviews of unconventional technical processes and biological resources of potential importance to developing countries.